Geographic Information Systems and Applications in Coastal Studies

Edited by Yuanzhi Zhang and Qiuming Cheng

Published in London, United Kingdom

Geographic Information Systems and Applications in Coastal Studies
http://dx.doi.org/10.5772/intechopen.97909
Edited by Yuanzhi Zhang and Qiuming Cheng

Contributors
Amit Kumar Jamwal, Vikram Sharma, Dibyendu Dutta, Tanumi Kumar, Chiranjivi Jayaram, Wasim Akram, Rizwan Ahmed, Mujahid Ali Khan, Haris Hasan Khan, Ali Atef Yousef Ali Masria, Karim Nassar, Mohamed Galal Eltarabily, Mustapha Nassir, Arati Paul, Chandra Shekhar Jha, Abhinav Galodha, Chander Prakash, Devansh Raniwala, Elsai Mati Asefa, Kefelegn Bayu Barasa, Dechasa Adare Mengistu, Parmita Ghosh, Siva P. Kumpatla, C. Prakasam, R. Saravanan, Arbnor Pajaziti, Orlat Tafilaj, Charles Bwalya Chisanga, Chizumba C. Shepande, Edson Nkonde

First published in London, United Kingdom, 2022 by IntechOpen
IntechOpen is the global imprint of INTECHOPEN LIMITED, registered in England and Wales, registration number: 11086078, 5 Princes Gate Court, London, SW7 2QJ, United Kingdom

British Library Cataloguing-in-Publication Data
A catalogue record for this book is available from the British Library

Additional hard and PDF copies can be obtained from orders@intechopen.com

Geographic Information Systems and Applications in Coastal Studies
Edited by Yuanzhi Zhang and Qiuming Cheng
p. cm.
Print ISBN 978-1-80355-741-0
Online ISBN 978-1-80355-742-7
eBook (PDF) ISBN 978-1-80355-743-4

For EU product safety concerns:
IN TECH d.o.o., Prolaz Marije Krucifikse Kozulić 3, 51000 Rijeka, Croatia,
info@intechopen.com or visit our website at intechopen.com.

Contents

Preface

This book provides a comprehensive overview of techniques, approaches, and experiences in the rapidly growing field of geographical information systems (GIS) in coastal zones. It focuses on advances in GIS technology and applications. Most of the work represented in this book is based on innovative applications of geographic information in coastal environments and disasters, coastal resources, coastal social systems, and coastal urban environments as well as applications of new algorithms, big data processing, and deep learning at global, regional, or local scales. Specifically, the chapters discuss GIS in coastal asset mapping and management, shoreline change analysis, development of ancient cities, agriculture and land suitability analysis, solid waste management, impact of COVID-19 measures on air quality, dynamic modelling of LULC, strategic planning of hydroelectric projects, groundwater potential zone identification, and road optimization for effective management traffic and transport. This volume is a useful resource for those who have used GIS for monitoring, modeling, planning, and policymaking in earth sciences and coastal studies.

We thank all the contributing authors and reviewers for their excellent work, helpful comments, and timely feedback. We are also grateful for the guidance and support from Publishing Process Manager Ms. Elena Vracaric and the editorial board of IntechOpen. Some of the in situ observations and field data used in some chapters were obtained from projects at the National Natural Science Foundation of China (U1901215) and the Marine Special Program of Jiangsu Province in China (JSZRHYKJ202007).

Dr. Yuanzhi Zhang
Nanjing University of Information Science and Technology,
School of Marine Sciences,
Nanjing, China

Dr. Qiuming Cheng
Sun Yat-sen University,
School of Earth Sciences and Engineering,
Zhuhai, China

Section 1

Geospatial Technology and Application

Chapter 1

Open-Source Geospatial Technology for Coastal Asset Mapping and Management

Arati Paul, Dibyendu Dutta and Chandra Shekhar Jha

Abstract

Coastal areas are highly productive in terms of natural resources and important for their strategic location. This leads to the development of industry, ports, and townships/cities in coastal areas. Therefore, many assets/ infrastructures are developed in coastal areas which are managed by respective administrations. Proper management of these assets requires a geographic information system (GIS) that can integrate and disseminate geospatial data along with ground photographs. The advancement of open-source geospatial technology enables the development of feature-rich yet cost-effective GIS applications. In the present chapter, one such web GIS framework is discussed for mapping and management of coastal assets using open-source geospatial technology.

Keywords: coastal, asset, mapping, geospatial technology, open source

1. Introduction

Coastal areas are bordering areas of a land surface that are close to a coastline or seashore where land and water surfaces meet. They are unique areas in terms of biodiversity and ecosystems. Coastal areas are environmentally sensitive and valuable with respect to the economy. They are extremely important to human activities, such as settlements, ports, business, and lively hood. Coastal fisheries are one of the major economic activities in coastal areas other than tourism and industry. It also provides opportunities for the generation of renewable energy. Therefore, many assets/infrastructures are developed in coastal areas which are managed by respective administrations. Coastal ecosystems provide protection against sea-level rise and tsunamis. However, natural calamities, such as cyclones and floods affect coastal regions frequently impose threats to economic and environmental assets. Proper management of these assets requires a GIS that can integrate and disseminate geospatial data along with ground photographs. The advancement of open-source geospatial technology enables development of feature-rich yet cost-effective geographic information system applications using free and open-source software (FOSS). Consequently, geospatial technology using FOSS is increasingly utilized in different sectors viz. natural resource management [1], biodiversity conservation [2], plantation and green space management [3, 4], planning of infrastructure [5]. In the present chapter, one such web GIS framework is discussed for mapping and management of coastal assets using open-source geospatial technology, including mobile apps.

Mapping of the assets is done through a field survey. GPS-enabled devices are used for collecting the spatial location of the asset. The smartphone-based geotagging application not only collects the location but also captures the asset photograph along with other associated information. The open-source Android operating system (OS) provides an excellent application programming interface (API) for developing such geotagging app that helps in systematic mapping of coastal assets and automatic generation of asset database. In the present chapter, the development process of a geotagging app using Android Studio is discussed. Once the asset data is captured in the field, the database is created. Open-source Postgres database application along with PostGIS extension provides support for spatial database management. PHP is an open-source scripting language that interfaces the database to retrieve information and passes them to the client application. The client application developed using HyperText Markup Language (HTML), JavaScript, etc. enables in generating dynamic web pages to send and receive user queries and replies respectively from the server. In the present chapter, the development framework of a WebGIS application involving open-source geospatial technologies is discussed.

A brief introduction of WebGIS, including its characteristics and components, is given in Section 2. Section 3 introduces mobile applications in the context of location-based services. The methodology is described in Section 4 whereas the results are discussed in Section 5. Finally, conclusions are drawn in Section 6.

2. WebGIS

GIS system includes computer hardware, software, and applications to capture, edit, analyze, manipulate, and visualize geo-referenced data. WebGIS is an advanced form of GIS available on the web platform where the exchange of information takes place between a GIS server and a client (mostly browser-based) application running on a mobile or desktop.

2.1 Characteristics of WebGIS

A diverse set of analytical functions are offered by GIS beyond mapping. WebGIS extends the potential of GIS to a wider audience and helps in the decision-making. Some of the important characteristics of WebGIS are listed below:

- It serves a large number of users simultaneously
- Runs on browser and supports multiple platforms and operating systems.
- Follows unified updates.
- Easy access to map and satellite imagery for different applications.
- Available in customized dashboards with user-friendly interfaces.

2.2 Components of WebGIS

WebGIS allows the dissemination and analysis of geospatial data over the web. Components of WebGIS include (i) database server, (ii) GIS server, (iii) application server, and (iv) client. **Figure 1** shows the architecture of WebGIS where each component is connected. It is a three-tier architecture. The first and second tires include the client (Browser) and application server respectively, whereas the GIS server and the database server reside in the third tier. The communication between these components is performed via Hypertext Transfer Protocol (HTTP) and the format of the response can be an HTML, binary image, XML (Extensible Markup Language), GML (Geography Markup Language), or JSON (JavaScript Object Notation).

The client accesses the WebGIS application hosted in the application server through HTTP requests. The WebGIS application provides a customized interface that includes functionalities viz. display, overlay, query, and analysis. The client requests are analyzed in the application server and accordingly, requests are sent to either/both GIS server or/and database server. On a successful query, responses are sent to the application server from the database and/or GIS server. Finally, results are presented to the client in customized form by the application server through the HTTP response. Brief functionalities of each of the WebGIS components are described in the following subsections.

2.2.1 Database server

A GIS includes spatial and aspatial data which are stored in a database. The database is suitable for large datasets with several features and provides an efficient mechanism to store, query, analyze, and update these data [6]. Database server communicates with the GIS server and application server for providing spatial and aspatial data as per their requests.

2.2.2 GIS server

The GIS server is software that creates web services using spatial data for GIS applications. Web service runs on the GIS server and performs some actions in response to a client request. To facilitate the exchange of geographic data across the

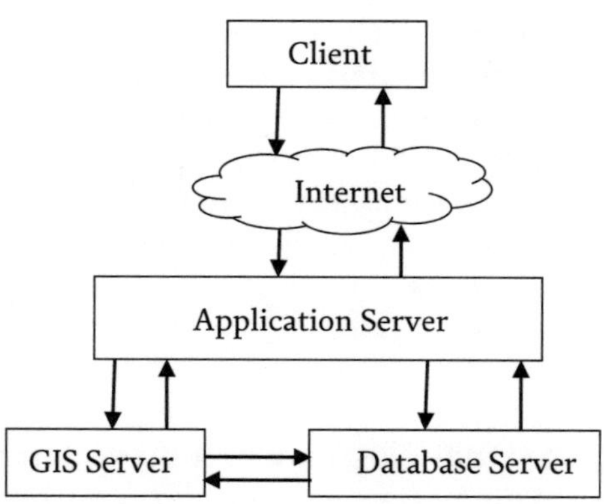

Figure 1.
WebGIS architecture.

web, the Open Geospatial Consortium (OGC) has defined open specifications for GIS web services. Different OGC compliant services include the web map service (WMS) and web map tile service (WMTS) for requesting maps, the web feature service (WFS) for requesting vector feature geometries and attributes, the web coverage service (WCS) for raster data requests, and the web processing service (WPS) for spatial data processing operations.

2.2.3 Application server

The application server hosts the WebGIS application that accepts clients' requests and processes them and provides the output in customized form. Different APIs and libraries viz. OpenLayers and Leaflet are used to develop the WebGIS application that consumes web services. API is an organized set of programming components that can be used to develop applications using a particular language. The application handles the details of all the web service requests occurring in the background and provides an intuitive user experience. In WebGIS application maps are often produced by combining multiple web services as layers. The term "mashup" refers to describe such maps created from multiple web services [7].

2.2.4 Client

Client refers to a computer or other device that requests information or processing from a server. The client accesses the WebGIS application hosted in the application server through the browser via the Internet. Each time an action is performed at the client, the request is sent to the application server. On successful processing of the request, the result is displayed at the client end in a customized form in the WebGIS application.

In spite of having advantages over standalone GIS, WebGIS have several challenges viz. security, data quality, performance, poor network connectivity, and maintenance.

3. Mobile application for location-based services

The geographical location of an object along with its description is an important aspect of mapping. Objects are mapped in the field using traditional tools and methods that have their own limitations. In recent years, digital methods with new technologies are gaining significant interest and popularity for field surveys and mobile-based mapping is one of them. Smartphones and tablets integrated with Global Positioning System (GPS) receiver provide a convenient way of recording location as well description of observations. Hence, in the present section, a few important technological tools/functionalities are described briefly that help in developing mobile applications for location-based services.

3.1 GPS module

GPS, a satellite navigation technology, provides the geolocation information of an object and is mostly used in the location-based mobile app. The location accuracy of recent GPS technology is within a few meters.

3.2 Cellular ID

Cellular identification is unique for any device. It enables in getting the approximate location information of the device from the cellular tower. It is an essential component of geolocation in the absence of live data from the mobile device.

3.3 Geofencing

A geofence is a virtual boundary set up around a geographical location. Geofencing is a location-based service in which an app uses GPS or cellular data to trigger certain predefined actions when a mobile device enters or exits a geofence.

3.4 Map functionalities

The map is an important element that supports the display and analysis of location information. In addition to the display of map and satellite view, some significant functionalities of map elements include routing/navigation, geocoding, measurement of distance/ area, etc.

4. Methodology

Coastal asset mapping and management system involve two subsystems, the first one is mapping and the second one is management. In this section, the methodology for designing and developing the total system is discussed using open-source geospatial technology.

4.1 System architecture

The mapping of coastal asset is carried out using location-based services of the mobile device whereas the management of assets is implemented through WebGIS. The overall system architecture is given in **Figure 2**. The mobile application is developed to collect coastal asset data in the field. This data includes field photos as well as asset information. Once the data is collected by the app, it is saved on the device. Subsequently, the data is sent to the database server via the application server through the Internet. GIS server hosts the WMSs related to base map layers. The database server serves the spatial and aspatial data to the GIS server and application server depending on the requests. The development methodology of each of the components using open-source tools and software is described in the following subsections.

4.2 Design and Development of Mobile application

The most popular open-source mobile-operating system is Android. Hence, in the present chapter, the development of an Android-based mobile application is discussed using Android Studio. Android Studio is the official integrated development environment (IDE) for Android and contains features that are required to build an Android app. An IDE is software that combines common developer tools into a single graphical user interface (GUI) for building applications. Some important features of Android Studio include visual layout, APK analyzer, intelligent code editor, fast

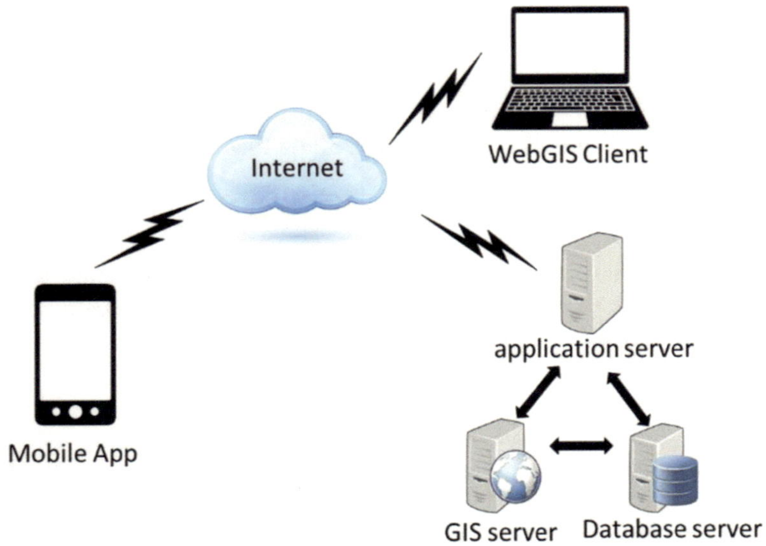

Figure 2.
Overall system architecture.

simulator, flexible build system, etc. that accelerates the development process and helps to develop quality applications for Android devices.

Since the mobile, as well as the WebGIS, shares a common database, the database is designed first. The main objective of the mobile app is to collect coastal asset information. Therefore, the asset table is created using the fields viz. asset name, asset type, locality, latitude, longitude, date/ time stamp, and collector's information.

The Java Programming Language, Android software development kit (SDK), and SQLite Database were used for developing the application. Java is a general-purpose programming language that is class-based, object-oriented, and designed to have as few implementation dependencies as possible. Android SDK, a collection of software development tools and programs, is used to develop new applications for devices running the Android operating system. Android apps can be developed using Kotlin, Java, and C++ languages using the Android SDK. The SQLite is an embedded SQL database engine that is used here for storing the collected survey data in the mobile device through the mobile app.

4.3 Design and development of WebGIS application

The Coastal asset WebGIS application may include user authentication to ensure secure data access. The user interface of the application is designed as a dashboard. The main application page contains a map with elements viz. legend, table of content, overview map, and scale bar. Different GIS functionalities viz. Navigation and Measurements are developed as different tools and combined under the "Tools" menu. All the tools and map elements are made collapsible to facilitate effective map visualization.

4.3.1 Preparation of map layers

Coastal asset mapping and management is a special kind of GIS application. The base map may include layers viz. road and rail network, population information, land use/land cover (LU/LC), administrative boundaries, drainage, and water bodies. Remote sensing and GIS techniques are utilized for the generation of various thematic resource maps in conjunction with ancillary data. Quantum GIS (QGIS) [8] is a free desktop GIS software that can be used to generate these spatial layers. QGIS supports both raster and vector layers allowing users to analyze them for spatial map generation. External WMS and WFS are also supported in QGIS [9].

Remote sensing satellite data, such as CARTOSAT and LISS-IV, may be integrated with the GIS environment to extract information related to natural resources and generate thematic layers. These utilitarian types of maps serve the purpose of planning and decisions making.

4.3.2 Generation of database server

Since the mobile and WebGIS application shares a common database the design of the database should be performed at the beginning before the actual development starts. The structure of the coastal asset data table is generated along with its field definitions and data types. This table is used to save the asset records captured through the mobile app. The same table is used by the WebGIS application for the display and analysis of coastal assets.

A spatial database is required to store and utilize the asset data in the web environment. PostgreSQL/PostGIS is a powerful reliable and stable open-source database [10, 11] and is used for this purpose. It supports most of the major operating systems viz. Linux, UNIX, and Windows for managing spatial/aspatial data. PostGIS extension of PostgreSQL provides support for geographic objects. This enables the performance of spatial queries in PostgreSQL. Hence, the coastal asset database is created in PostgreSQL/PostGIS where the asset table is generated. Subsequently, the other geospatial layers are incorporated in the coastal asset database one after another.

4.3.3 Formation WMS

Once the database is prepared, the spatial layers are converted into Web Map Service (WMS), to facilitate spatial data sharing over the Internet. The open-source Java-based GeoServer [12] provides the platform to share, edit, and display geospatial content in the web environment. It accesses data from any major spatial data source viz. PostGIS, ArcSDE, Oracle, and DB2 and publishes those using open standards. GeoServer enables very quick and easy map generation using the free java-based mapping library "OpenLayers."

4.3.4 Development of client web application

The interactive WebGIS application for the management of coastal assets is developed using HTML and Java scripts, while OpenLayers [13] mapping library is used to develop geospatial functionalities. OpenLayers enables the development of feature-rich WebGIS applications by putting a dynamic map on the web page [14]. Tiled layers can be rendered using OpenLayers from OGC complied sources viz. open street map (OSM), Bing, and MapBox. A range of vector data formats, including GeoJSON, TopoJSON, KML, and GML, is supported by OpenLayers. Operations, such as map rendering, interactive drawing, and editing can be performed by OpenLayers without using an additional plug-in at the client end.

PHP (Hypertext Preprocessor) [15] is used to query the database and fetch results as per the user's requirement. PHP is a fast and flexible general-purpose scripting language that can be easily embedded in an HTML page. Hence, it is very popular and useful in web-based application development. It is used as server-side script to interface the database from the WebGIS application.

The query results fetched from the database are dynamically managed using AJAX (Asynchronous JavaScript and XML) [16] web development technique. It asynchronously sends the request and retrieves data from the server in the background without reloading the web page. To achieve this, it uses a group of existing web technologies together, including HTML, Cascading Style Sheets (CSS), JavaScript, Document Object Model (DOM), XML, and the XMLHttpRequest object. The dynamic results are parsed and analyzed using the jQuery JavaScript library [17]. The feature-rich jQuery library simplifies event handling and AJAX in the WebGIS application. CSS dictates the appearance of HTML elements of a page and hence is used to develop the look and feel of the user interface of the present application.

5. Results and discussions

The entire system contains two major applications one is the mobile app for coastal asset data collection and another is WebGIS application for data visualization and management. The functionalities of the asset data collection app are described in section 5.1 while the WebGIS functionalities are given in Section 5.2.

5.1 Mobile app for coastal asset data collection

The mobile application starts with a user profile screen. It takes the user's information viz. name, email id, and phone number, and validates them (**Figure 3a**). Information once entered is stored for subsequent use, however, at the beginning of the next use, the user can update them. Upon receiving the valid user information, the app proceeds to the main activity (**Figure 3b**). Here, options are provided to input information and take a picture related to coastal assets. The current latitude and longitude positions are captured by the GPS receiver of the mobile phone. Finally, the record consisting of all these data can be sent to the server along with the current date and time stamp using the "Send" button. The captured information can be saved in the local device using the "Save" button. The Exit button enables exiting the application. The "Records" button enables viewing and editing of saved records. The mobile app also enables the user to visualize the sent records using the "Map" button.

5.2 WebGIS functionalities

Once the captured record is sent to the centralized server, the web application displays those records on the map. The display of the map is shown in **Figure 4** where the coastal area of the state of West Bengal, India is shown. The Bhuvan [18] WMS is used for the display of the base map and satellite imagery. The satellite view shows the high-resolution satellite image of the area.

The map also shows the scale bar as well as the overview map at the bottom left and the bottom right corner, respectively. Here three asset locations are plotted using blue markers which were collected using the mobile app. Different markers/colors

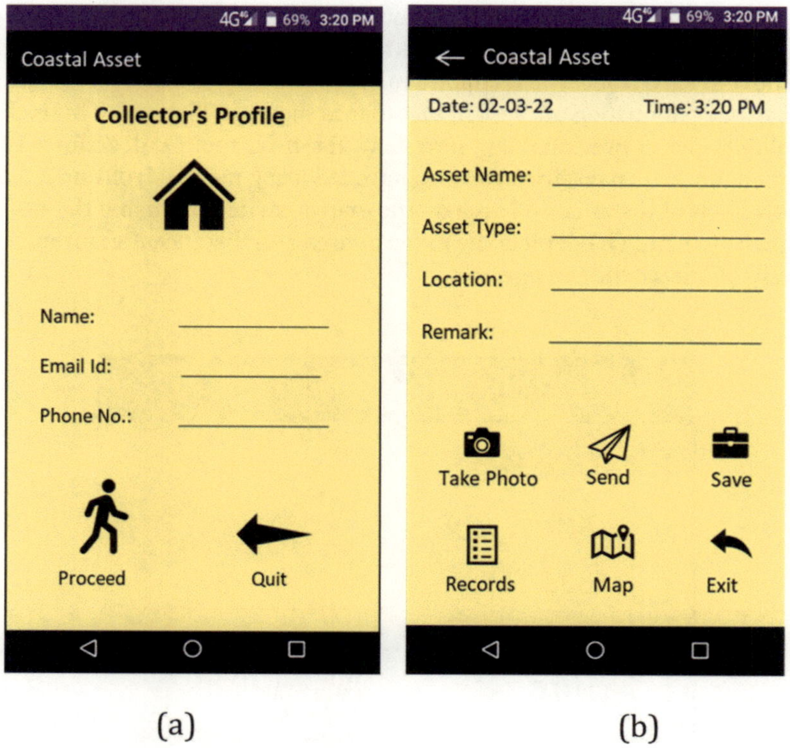

Figure 3.
Mobile app (a) home screen, and (b) main activity screen.

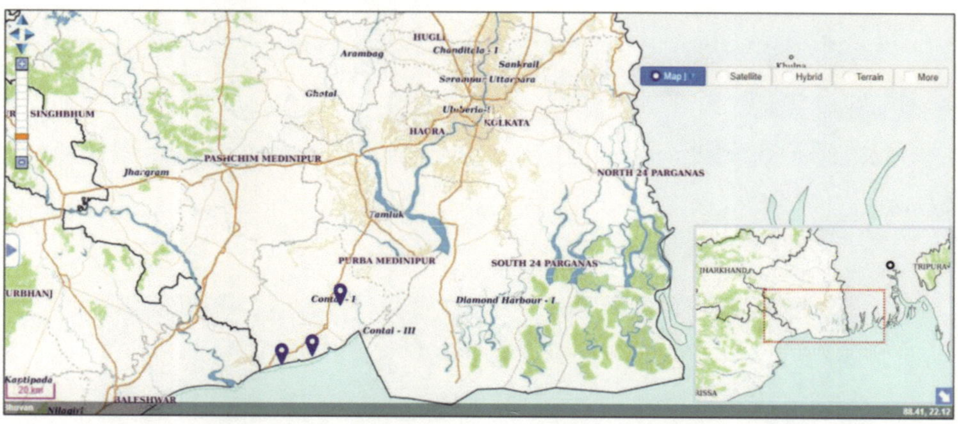

Figure 4.
Map display.

may be assigned to display different types of coastal assets on the map viz. administrative, financial, educational, industrial along with the proper legend.

Another important functionality of the WebGIS is overlay where individual map layers (here WMS) are overlaid on the base map. Users can control the overlay through the content window that contains all the spatial layers available in the application. **Figure 5** shows a sample content window that enlists administrative boundaries viz. block boundary, district boundary, road, drainage, land use/land cover. Users can add or remove these layers as and when required. The zoom-in button is placed next to each layer which enables to zoom the map to the layer extent.

The map navigation tools are kept in a toolbar as shown in **Figure 6**. This includes functionalities, such as previous view, next view, zoom-in, zoom-out, zoom to full extent. In addition to this map, navigation can be controlled using mouse scroll and left buttons.

Measurement of distance and area on the map is carried out using the respective tools. **Figure 7(a)** and **(b)** depicts the usage of area and distance measurement tools respectively on the satellite image.

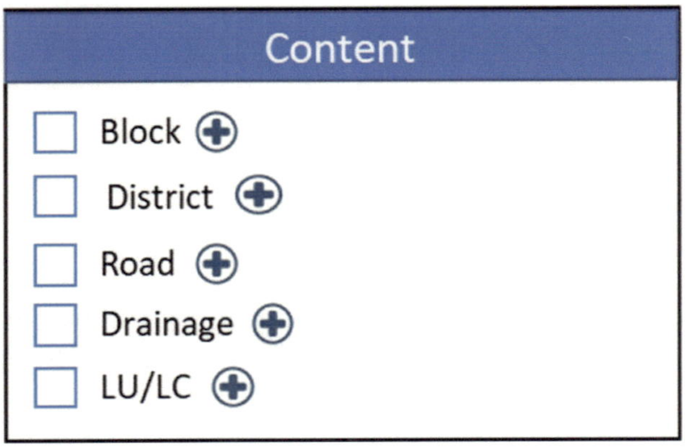

Figure 5.
Map overlay with content window.

Figure 6.
Map navigation tools.

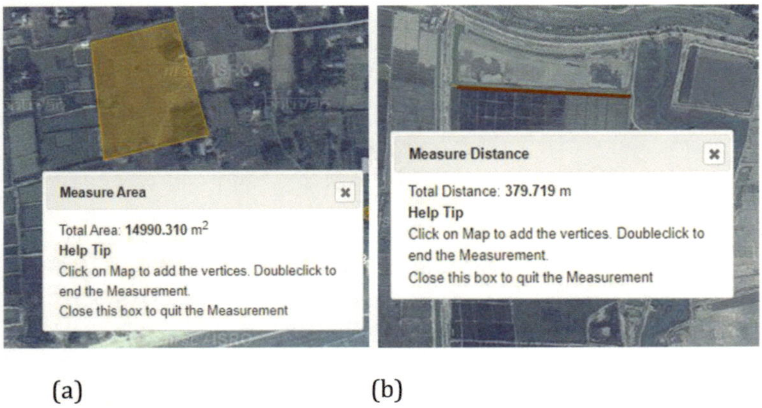

Figure 7.
Measurement tool for (a) area, and (b) distance.

Search is an important functionality of the GIS application. Search on the coastal asset layer can be performed by type/name/location. Search results in the display of assets that matches the criteria along with statistics. The output can be further displayed using charts. Spatial as well as aspatial queries also can be implemented depending upon the user's requirements. The proximity analysis enables to identify assets in a given buffer distance of a selected location in the map. In addition to it, the user can select a particular category of the asset to the given proximity of a location. **Figure 8(a)** shows the location of health facilities (by the red marker) available in the 5 km proximity from the location denoted by the black marker. The proximity tool is shown in **Figure 8(b)**. The map "onclick" functionality shows the pop-up window to provide further details of an asset along with the ground photograph as given

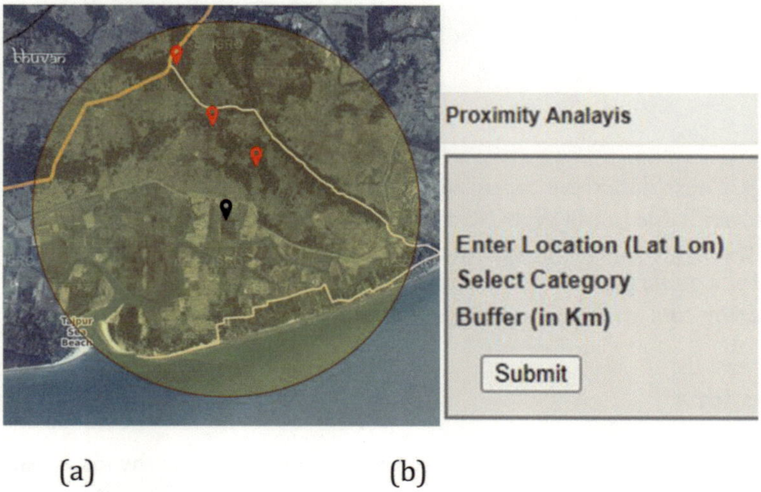

Figure 8.
Proximity analysis.

Figure 9.
Map 'on-click'.

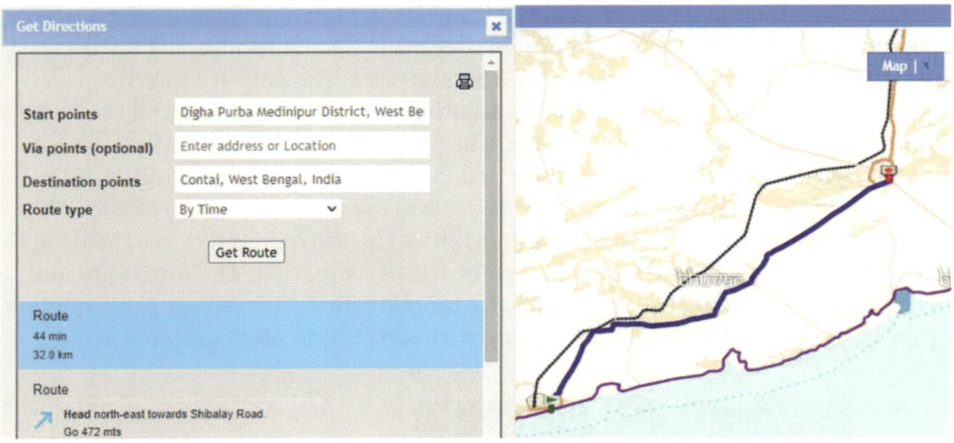

Figure 10.
Route direction.

in **Figure 9**. These data were captured and sent through the mobile app and subsequently are available in the WebGIS application in near real time.

Routing comes under network analysis where the road network is used to get route direction for a given pair of source and destination locations chosen by the user. The route direction tool and the map output are shown in **Figure 10**.

6. Conclusions

In recent years, geospatial technology is effectively utilized in the management of natural resources and spatial infrastructures [19, 20]. Compared to conventional methods, this technology is efficient and hence is widely adopted in different applications across the world. It has become more popular and affordable among the stakeholders due to the availability of FOSS. In the present chapter, an open-source framework is discussed for coastal asset mapping and monitoring using geospatial technology. This enables fast capture and analysis of coastal asset information in near real time in a cost-effective manner. Mobile app-based asset data collection involves minimum human interaction that leads to more accurate data entry and less error. The system provides authenticate and reliable data access mechanism. The WebGIS application is accessible by a large number of stakeholders without any financial investment at the individual level except the Internet connectivity. It has great customization capability [21] to encode the complex business logic that enables effective decision-making even by non-GIS professionals.

Acknowledgements

The authors acknowledge the Bhuvan portal and the team for providing supporting map functionalities.

Conflict of interest

The authors declare no conflict of interest.

Other declarations

Results are demonstrated using synthetic data.

Author details

Arati Paul[1*], Dibyendu Dutta[1] and Chandra Shekhar Jha[2]

1 Regional Remote Sensing Centre—East, NRSC, ISRO, Kolkata, India

2 Regional Centres, NRSC, ISRO, Hyderabad, India

*Address all correspondence to: aratipaul@yahoo.com

References

[1] Paul A, Chowdary VM. Open Source Geospatial Technologies for Generation of Water Resource Development Plan, Mapping, Monitoring and Modeling Land and Water Resources: Advanced Techniques Using Open Source Software. Boca Raton: CRC Press; 2021. DOI: 10.1201/9781003181293-12. ISBN: 9780367486839

[2] Mose VN, Western D, Tyrrell P. Application of open source tools for biodiversity conservation and natural resource management in East Africa. Ecological Informatics. 2018;**47**:35-44. DOI: 10.1016/j.ecoinf.2017.09.006

[3] Chakraborty D, Paul A, Dutta D, et al. Indian tea garden information system—A WebGIS enabled solution. Asian Journal of Geoinformatics. 2016;**16**(1):8-16

[4] Paul A, Mondal RS, Chakraborty D. Open source geospatial solution for disseminating green park information. Journal of Information Technology and Computer Science. 2019;**4**(1):57-63. DOI: 10.25126/jitecs.20194194

[5] Paul A, Mal P, Gulgulia PK, et al. Spatial progression of estate property management system with customized freeware GIS. International Journal of Information Technology. 2019;**11**(2):341-344. DOI: 10.1007/s41870-018-0135-y

[6] Mondal RS, Chakraborty D, Paul A, Dafadar KD. WebGIS enabled facility mapping and identification—A cost effective solution. International Journal of Computer Applications. 2018;**180**(38): 41-44. DOI: 10.5120/ijca2018917019

[7] Quinn S. In: Wilson JP, editor. Web GIS. The Geographic Information Science & Technology Body of Knowledge (1st Quarter 2018 Edition). 2018. DOI: 10.22224/gistbok/2018.1.11

[8] Menke K, Smith R Jr, Pirelli L, Hoesen JV. Mastering QGIS. Birmingham: Packt Publishing; 2015. ISBN 978-1-78439-868-2

[9] Cavallini P. Free GIS Desktop and Analyses: QuantumGIS, the Easy Way. The Global Geospatial Magazine. 2007

[10] Douglas K, Douglas S. PostgreSQL—A Comprehensive Guide to Building, Programming and Administering PostgreSQL Databases. Sams Publishing; 2003

[11] Kraft TJ, Mather SV, Corti P, et al. PostGIS Cookbook. 2nd ed. Birmingham: Packt Publishing. ISBN: 9781788299329; 2018

[12] Iacovella S. GeoServer Cookbook. 1st ed. Birmingham: Packt Publishing. ISBN 978-1-78328-961-5; 2014

[13] Santiago A. The Book of OpenLayers 3. Lean Publishing; 2014

[14] Hazzard E. OpenLayers 2.10: Beginner's Guide. Birmingham: Packt Publishing; 2011

[15] Lurig M. PHP Reference: Beginner to Intermediate PHP5. ISBN: 978-1-4357-1590-5. 1st ed., 2008. Lulu.com

[16] Darie C, Brinzarea B, Chereches-Tosa F, Bucica M. AJAX and PHP. 1st ed. Birmingham: PACKT Publishing; 2006

[17] Chaffer J, Swedberg K. Learning jQuery. 4th ed. Birmingham: Packt Publishing. ISBN 978-1-78216-314-5; 2013

[18] Bhuvan [Internet]. 2022. Available from: https://bhuvan.nrsc.gov.in [Accessed: March 4, 2022]

[19] Paul A, Chowdary VM. Application of web enabled open source geospatial technologies in generation of water resource development plan. International Journal of Hydrology Science and Technology. 2021;**11**(1):76-87. DOI: 10.1504/IJHST.2020.10023542

[20] Paul A, Chakraborty D, Dutta D, Das PK, Jalan M, Chettri S, et al. Park information system for Kolkata- A low cost web enabled solution. Journal of current trends in information technology. 2016;**6**(1):1-5

[21] Paul A, Chowdary VM, Chakraborty D, Dutta D, Sharma JR. Customization of Freeware GIS Software for Management of Natural Resource Data for Developmental Planning- A Case Study. International Journal of Open Information Technologies. 2014;**2**(4):25-29

Section 2

Shoreline Change Analysis and Assessment

Chapter 2

Shoreline Change Analysis of Hooghly Estuary Using Multi-Temporal Landsat Data and Digital Shoreline Analysis System

Dibyendu Dutta, Tanumi Kumar, Chiranjivi Jayaram and Wasim Akram

Abstract

Long-term (1973–2021) shoreline displacement, rate of change, and temporal pattern were examined using multi-date Landsat data and Digital Shoreline Analysis System (DSAS) along the 200 km coast of Hooghly estuary. Orthogonal transects of 100 m apart were casted for calculation of End Point Rate and Weighted Linear Regression rate on different temporal scales for seven analysis zones. The shoreline change pattern was established using Hierarchical agglomerative clustering. The study reveals that almost 43.45% of the beachfront has eroded and 56.55% has accreted during the past four decades. The average erosion rate varies between −0.01 and −13.71 m $yr.^{-1}$ and accretion of −0.01 to −22.30 m $yr.^{-1}$. The littoral drift resulted in a maximum seaward aggression by 1096.89 m in the zone 1. Landward movement was maximum (−602.96 m) in the zone 4. Although west bank is prograding @ 3.47 m $yr.^{-1}$ (±5.83), the east bank is eroding @ 1.30 m $yr.^{-1}$ (±4.08). Based on the cluster analysis about –1.87% of the shoreline exhibits consistent erosion over all the intervals, whereas trend was evident in 4.73% of the coastline. The portions of coastlines, which exhibit high erosion rate and consistent erosion need immediate attention and policy intervention.

Keywords: Hooghly estuary, shoreline change, erosion, accretion, Landsat, DSAS, change pattern

1. Introduction

The shoreline is the physical interface or intertidal margin between land and sea and constitutes one of the 27 global "Geo-indicators" referred by the International

Union of Geological Science [1] and International Geographic Data Committee (IGDC). Shoreline change is a dynamic natural process in the coastal areas induced by erosion/accretion that occurs over a range of temporal scales. The morphological evolution of the Hooghly estuary and its coastline is the result of two counteracting transport processes of sediment supply versus removal. When both the processes are balanced an equilibrium is reached. However, most often this balance is disturbed due to the influence of episodic and/or long-term natural forcing and anthropogenic interventions. As a consequence, the shoreline keeps changing its position [2–6] over a wide temporal scale, from geologic age to short-lived, extreme weather events such as storms and tsunamis. The long-term processes that shape the shoreline include sea-level rise (SLR), altered wind patterns [7], frequency and intensity of storms [7], offshore bathymetric changes [8], high energy swells [9] and supply of fluvial sediment input. In addition, anthropogenic activities *viz.*, landcover changes in the river catchment, port and harbor and dam construction, dredging for maintaining navigation channels, aquaculture, protective embankments, beach nourishment, economic and tourist activities also exacerbate the coastline change on a short temporal scale. Engineering structures change the estuarine circulation patterns and may change the freshwater flow along with sediment and nutrient supply. In several instances, engineering modifications to the beach creates discontinuities in the historical shoreline position and mask underlying long-term behavior [10]. Another less reported phenomena are land subsidence which may occur naturally due to compaction of sediments or triggered by the excessive withdrawals of ground water. In general, the coastal landform establishes a morphodynamic equilibrium after episodic short-term perturbations. However, many times the combination of natural and manmade activities exacerbates the shoreline change and exhibits non-linear morphological responses to change [11].

According to Williams [12], the study of shoreline variation and forecast plays an important role in coastal zone management and it becomes more crucial in the context of anticipated climate change and sea-level rise [13]. In this context, one of the key requirements for effective coastal zone management is the availability of accurate position of the shorelines for analysis of changes in the past and future trends. Traditional methods of shoreline delineation include terrestrial surveys using landmarks, aerial photos [14, 15], Global Positioning Systems (GPS), terrestrial Light Detection and Ranging (LiDAR) or 3D scanners. But they are time-consuming, labour intensive and costly. In contrast the remote sensing data form space platform is more convenient, easy to process and above all freely available in the public domain. Remote sensing data has been extensively used in shoreline change studies because of their synoptic and repetitive coverage, multispectral capabilities enabling contrast between land and water in the infrared portion, and cost-effectiveness [14, 16]. Advanced image processing techniques can be employed on satellite data for precise extraction of the shoreline. Some of the methods used by different researchers include threshold level slicing and image classification technique [17], density slicing of TM band 5 [18], canny edge detection using DN threshold [19], mean shift segmentation [20], pixel-based segmentation using DN threshold [21], neural network [22], fuzzy logic [23, 24], texture analysis [25], machine learning [26] and incorporation of ancillary spatial data in the classification scheme [27–29]. Quantitative assessment of the spatio-temporal variation of shoreline at global scale has been carried out by several authors [30–32]. In this endeavor the twin technologies of Remote Sensing and Geographic Information System has been recognized as the most useful tools for quantifying the historic shoreline change [33, 34]. To avoid the discrepancy which might be introduced due to fluctuation of water level Yu et al. [35]

have used satellite images obtained at similar tidal heights. Chen and Chang [36] have done the tidal correction using high spatial resolution satellite images and real-time data of tidal level to reduce the impact of tidal level variability on the estimation of coastline change. In India also several studies have been carried out for shoreline change analysis using remote sensing data [37–39]. Most of the studies have used Digital Shoreline Analysis System [40], a software extension within the ArcGIS tool for measuring, quantifying, calculating and estimating of rate of change from multiple historic shoreline positions at different temporal scales [41–44]. The change metrics of DSAS are Net Shoreline Movement (NSM), Shoreline Change Envelope (SCE), End Point Rate (EPR), Linear Regression Rate (LRR) and Weighted Linear Regression Rate (WLR) among others. LRR and WLR enable multiple historic shorelines to be used to determine the rate of change by fitting a least-square regression line to all shoreline points for particular transects.

In the present study, Landsat satellite data of 8 temporal intervals between 1973 and 2021 were used for land-water discrimination, generation of shorelines and long-term change rate along with change pattern along the Hooghly estuary. The instantaneous land-water boundary was used as coastline which is relatively simple and can easily be identified using image transformation. The main objectives of the study are i) medium- and long-term changes in the shoreline at high spatial resolution using DSAS ii) to identify the erosion/accretion pattern and iii) to examine the role of change drivers.

The findings of the study will be useful for the managers and engineers to make scientific and rational policies for land use planning, to develop effective coastal protection strategies, predicting capacity for future coastal change due to climate and other drivers and improving impact and vulnerability assessments that include natural human sub-system interactions.

2. Study area

The Hooghly estuary is located in the southernmost part of Indo-Gangetic plain, flanked between East Midnapur (in the West) and South 24 Parganas district (in the East), extending between 21°33′10″N to 22°13′16″N latitude and 87°45′00″ to 88°18′22″E longitude (**Figure 1**). The head Bay is a unique deltaic environment comprising a wide continental shelf, complex coastal geometry and high tidal range. Tide domination is indicated by exponentially tapering channels, with funnel-shaped mouths [45]. The region has formed, sculptured and modified due to continuous fluvial action of the Ganga and the Brahmaputra systems, intense tidal hydrodynamic behavior, climatic disturbances and anthropogenic activities [46]. The funnel-shaped estuary has a width of 6 km at its head and 25 km at the mouth, responsible for tidal asymmetry and flow variation leading to bank erosion [47]. The average depth of the water column is only 6 m [48]. The estuary receives 4 tributaries *viz*. Damodar and Rupnarayan rivers at its head, and Haldi and Rasulpur rivers at the middle at its west bank. In contrast, the east coast is punctuated by several closely spaced inlets. Based upon the tidal amplitude, the coastal region of West Bengal can be sub-divided into i) macro tidal (tidal range > 4 m) from Sagar to Bangladesh border and ii) meso-tidal (tidal range of 2–4 m) mostly Medinipur (Digha-Sankarpur-Junut) coastal plains to the west of the Hooghly estuary.

Geologically the basement of the Bengal basin is a part of the eastern edge of the Indian plate, which is being subducted beneath the China plate along the Sunda

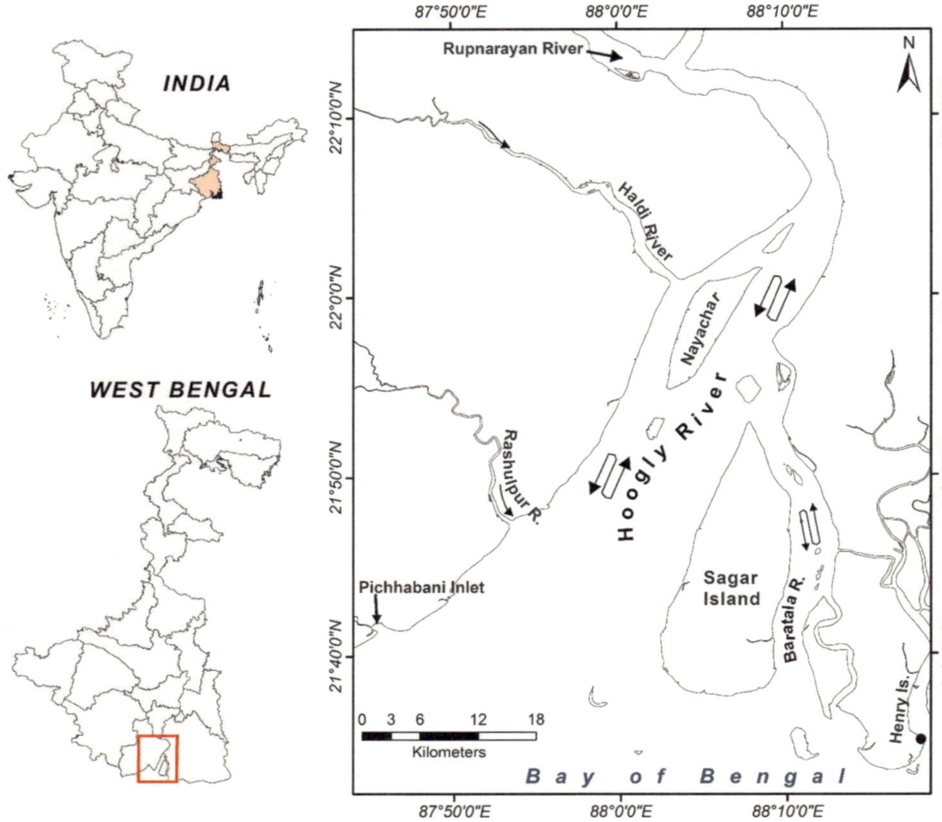

Figure 1.
Index map of the Hooghly estuary.

subduction zone and Naga-Lushai orogenic belt. The tectonic and depositional history of the Bengal basin has been controlled by several movements during Cretaceous-Tertiary periods. Due to the tectonic activity the Bengal basin has been tilted towards east resulting in successive changes in the course of the Ganga River towards east from the historical past. Due to this shifting, the deltaic region suffers from the paucity of fresh water discharge and sedimentation. Auto compaction of loosely attached sediments and gradual land subsidence is also another prominent geomorphic event occurring in this region [49–54] which mostly remains unnoticed. Morphometrically the Hooghly estuary is the product of continuous fluvial sedimentation in a series of para-deltaic lobe progradation systems developed on the western shelf margin areas and eastern troughs of the Bengal basin caused by the eustatic, isostatic and tectonic forces. The coastline presents various landforms such as tidal/mud flats, sandy beaches (located near Digha, Duttapur, Shyampur, Dadanpatra, Baguranjalpai, Dariapur and Nij Kasba), salt marshes (near Khejuri and at the mouth of Rasulpur river near Nij Kasba) and mangrove marsh (south of Patibunia). A vast extension of the muddy beach is found in South 24 Parganas, especially to the east of Bakkhali. The most striking feature is the development of successive rows of dunes (both Palaeo and Neo dunes) with intervening clayey tidal flats in the south of East Midnapur district between the stretches of Subarnarekha and Hooghly estuary is due to punctuations in the regression of the sea during Holocene [55]. Banerjee and Sen [56] opined that the regression of sea along this coastal tract is around 6000-year BP which resulted in seaward shifting of shoreline and formation of Paleo-dunes. Accordingly, to Niyogi [57], six regular cycles of beach ridges alternating with a variable number of bars are visible in the area, which is indicative of the shifting of shorelines. According to Gaur and Vora [58], the shoreline position was 5–15 km inland from the present shoreline around 6000-year BP. The erosion and accretion patterns clearly show a continued geomorphic sculpturing of the Hooghly coast.

To capture the micro-level variability, alongshore is divided into 7 analysis zones (**Figure 2**) covering both the west and east bank. The zones in the west bank are delimited by the main inlets which are the freshwater sources, eventually draining into Bay of Bengal. The area delimitation of various zones, constituting transects and shoreline distances is given in **Table 1**. The west bank is divided into 3 zones whereas the east bank into 4 zones (**Table 1**). The total length of the coastline studied is 200 km of which 90 km on the western side and 110 km on the eastern side of the estuary. The studied coastline was divided into 1924 number of transects (T_n) separated by 100 m. The number of transects increases from west to east bank in the clockwise direction.

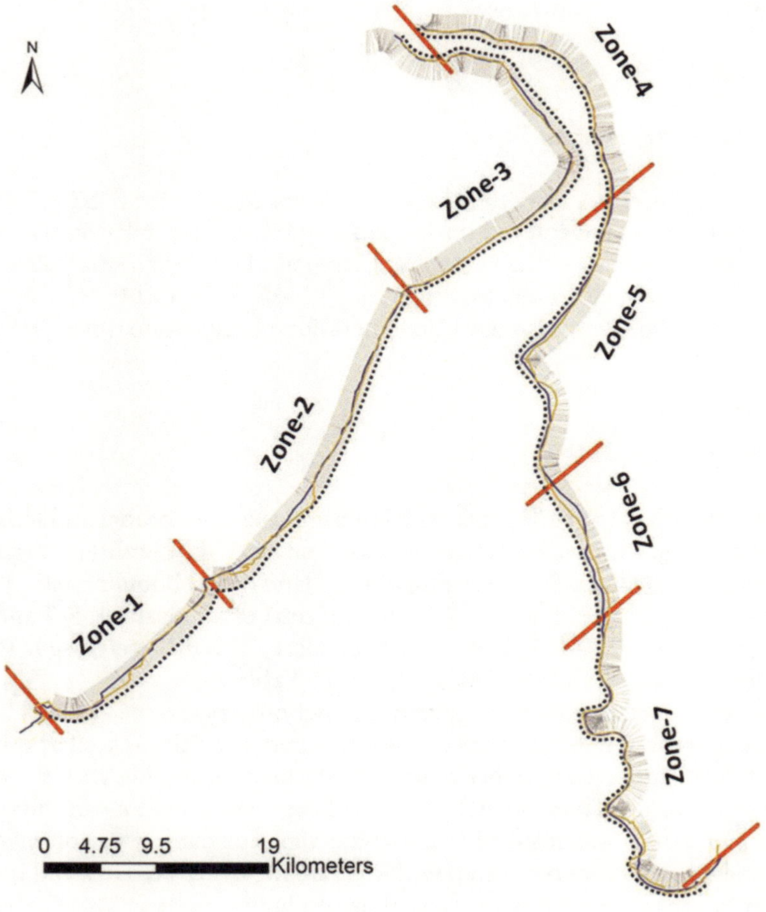

Figure 2.
Different analysis zones.

Zone	No of transects (from-to)	Location	Distance (km)
West bank			
Zone–1	187 (T_{25}-T_{211})	Pichhabani outlet to Rashulpur river	19.37
Zone–2	307 (T_{218}-T_{524})	Rashulpur river to Haldi river outlet	31.04
Zone–3	384 (T_{534}-T_{917})	Haldi river to the confluence of Rupnarayan and Hooghly River	39.52
East bank			
Zone–4	256 (T_{919}-T_{1174})	Confluence of Rupnarayan and Hooghly river to Kulpi	26.23
Zone–5	280 (T_{1175}-T_{1454})	Kulpi to Kakdwip	27.75
Zone–6	120 (T_{1455}-T_{1574})	Kakwip to Namkhana	13.41
Zone–7	390 (T_{1575}-T_{1967})	Namkhana to Henry Island	43.10

Table 1.
Salient description of different analysis zones.

3. Materials and methodology

3.1 Data used

3.1.1 Army and survey of India Topomaps

The historic shorelines were digitized from Army Map Series (NSS&H, Edition-1, AMS) in 1:250,000 scale surveyed during 1942–1943) number NF-45: 7 (north of study area) and 11 (south of the study area) were used for the coastline change analysis. Besides Survey of India topomaps of 73 N-16, 73O -13,14; 79B - 4; 79C-1,2,6 surveyed during 1967 were also used for generation of high-water level (HWL) coastlines.

3.1.2 Satellite data

Landsat satellite data of 1973 to 2021 have been used for decadal and long-term trend analyses. The data has been selected based on clear sky condition, high tide date and time as well as season. For discrimination of land-water boundary shortwave infrared bands 5 (1.55–1.75 μm) and 7 (2.08–2.35 μm) of Landsat - 4, 5, 7 and bands 6 (1.566–1.651 μm) and 7 (2.107–2.294 μm) of Landsat – 8 (OLI) were used. The details of the satellite data used in the study are given in **Table 2**.

For the of satellite data tide and current prediction programme *viz*. WXTide32 was used which allows knowing the time of high-tide and low-tide, as well as the tide height (m). It is supported by more than 9500 stations worldwide with the capability to predict tides from 1970 through 2037. One of the nearest stations of the study area, e.g., Diamond Harbor was selected to know the high tide date, time and magnitude. The high tide timings were compared with the satellite overpass dates and time for the selection of images representative of coastlines on high-tide date. This enabled comparison of the shorelines under identical tide conditions by minimizing the variability due to the tidal cycle.

Satellite/ sensor	Path/ Row	Date of overpass	Spatial resolution (m)	Overpass time (local time)	Time of high tide (local time)	Tide height (m)
MSS1	149/45	17.01.73	60	NA	09:40	4.05
MSS3	149/45	17.01.80	60	03:52	10:33	4.59
TM4	138/45	19.01.89	30	04:03	9:08	3.45
TM5	138/45	28.01.95	30	03:43	8:54	3.85
ETM + 7	138/45	06.03.00	30	04:23	11:15	4:73
TM5	138/45	07.01.05	30	04:17	8:14	3.81
ETM + 7	138/45	29.01.10	30	04:22	10:17	4.38
OLI 8	138/45	09.02.17	30	04:31	9:48	4.04
OLI 8	138/45	24.03.21	30	04:30	7:42	3.19

The tide information is pertaining to the Diamond Harbor station.

Table 2.
Details of the satellite data used.

3.2 Methodology

3.2.1 Land-water discrimination and shoreline extraction

There are seven types of coastline indicators *viz*. geomorphological reference lines, vegetation limits, instant tidal levels and wetting limits, tidal data, beach contours and storm lines. In the present study, the high-water levels (HWL) during spring tide have been considered for historical coastline change analysis. Several water indices have been used by many authors to extract the coastlines [59, 60]. Two of the most popular water indices are the Normalized Difference Water Index (NDWI) and Modified Normalized Difference Water Index (MNDWI). NDWI was primarily developed for Landsat MSS whereas MNDWI was developed for TM, ETM+ and OLI sensors [61]. The bands used to generate these indices mostly consist of green, near-infrared, middle-infra-red and shortwave infrared bands. The objective of all these indices is to enhance the contrast between land-water interfaces. Band 2 (0.52–0.60 μm) and band 5 (1.55–1.75 μm) of Landsat-7 and Band 3 (053–0.59 μm) and band 6 (1.57–1.65 μm) of Landsat-8 OLI were used for computing NDWI. The formulae of NDWI and MNDWI are given below:

$$\mathrm{NDWI_{MSS}} = \left(\mathrm{R_{green}} - \mathrm{R_{nir}}\right)/\left(\mathrm{R_{green}} + \mathrm{R_{nir}}\right) = (\mathrm{B_1} - \mathrm{B_4})/(\mathrm{B_1} + \mathrm{B_4}) \quad (1)$$

$$\mathrm{MNDWI_{TM,ETM+}} = \left(\mathrm{R_{green}} - \mathrm{R_{swir}}\right)/\left(\mathrm{R_{green}} + \mathrm{R_{swir}}\right) = (\mathrm{B_2} - \mathrm{B_5})/(\mathrm{B_2} + \mathrm{B_5}) \quad (2)$$

$$\mathrm{MNDWI_{OLI}} = \left(\mathrm{R_{green}} - \mathrm{R_{swir}}\right)/\left(\mathrm{R_{green}} + \mathrm{R_{swir}}\right) = (\mathrm{B_3} - \mathrm{B_6})/(\mathrm{B_3} + \mathrm{B_6}) \quad (3)$$

Where, R_{green} = spectral reflectance of the green band, R_{nir} = spectral reflectance of near-infrared band and R_{swir} = spectral reflectance of the shortwave infrared band.

Before applying the water index on Landsat MSS data of 1973, the image was resampled to 30 m spatial resolution to make the resolution comparable with the rest of the datasets. A Boolean approach was used on the NDWI/MNDWI images to create two classes *viz*. land and water. The threshold for land-water boundary was kept >0.15

for OLI and > 0.1 for TM/ETM+. The resulting image had only two classes *viz*. land and water. However, some of the inter-tidal zones could not be demarcated due to very high sediment loading. Hence, a hybrid approach was followed wherein NDWI/MNDWI, SWIR and topomaps were used for precise demarcation of the shorelines. Vectorization of the land-water boundary was done using the region growing tool of ERDAS/Imagine (ver. 9.1). The spectral Euclidean distance was set interactively to accurately capture the land water boundary. Some manual editing was also done on the shoreline vector.

3.2.2 Shoreline change analyses

Historical shoreline behavior was examined using Digital Shoreline Analysis System (DSAS, ver. 5.0), an extension tool of ArcGIS software (developed by the US Geological Survey) which calculates several change statistics *viz*. Net Shoreline Movement (NSM), Shoreline Change Envelope (SCE), End Point Rate (EPR), Linear Regression Rate (LRR) and Weighted Linear Regression Rate (WLR). It can analyze the time series of multiple shoreline positions [34] by using linear regression fit. Provision is there to include uncertainty of the input data in terms of assigned weights. Functionally DSAS performs 3 major activities *viz*., i) defining a baseline ii) generation of orthogonal transects and iii) computation of rates of changes. The tool enables the calculation of scales and rates of change statistics from multiple historic shoreline positions and sources. DSAS is a freely downloadable tool and is available at Woodshole [62] http://woodshole.er.usgs.gov/project-pages/dsas/. The details are available in Thieler and Danforth [63, 64] and Thieler et al. [34]. A brief description of the change statistics used in the study is given below.

Shoreline Change Envelope (SCE): A measure of the overall change in shoreline (m) at each transect considering the farthest and nearest position of the shoreline for the baseline location [65] irrespective of the dates.

Net Shoreline Movement (NSM): It is the distance (m) between the oldest and the youngest shorelines [66].

End Point Rate (EPR): It is derived by dividing the distance of shoreline change between two time periods by the time interval and expressed as m $yr.^{-1}$ [65–69]. This method provides the net rate of change over the long term. It has both advantages and disadvantages; the advantage is that only two shorelines are required for computation of change rate but unable to use more than two date shoreline data.

Linear Regression Rate (LRR): It determines a rate-of-change statistic by fitting a least square regression line using all the intersection points all shorelines and individual transect [66, 68, 70]. The slope of the line is the rate of shoreline change. The advantages of linear regression include i) all the time-series data are used and ii) can reduce the impact of spurious values on the overall accuracy of change rate [71]. To ensure meaningful results from the regression model the temporal intervals of shorelines were kept well distributed over the analysis period.

Weighted Linear Regression Rate (WLR): This method takes into account the positional and measurement uncertainty of the shoreline positions [63, 72–74] especially when the shorelines were digitized from various sources and scales. The uncertainty values are incorporated in the DSAS as weights. The slope of this regression line is the shoreline change rate (m $yr.^{-1}$). Using the weighing factor, WLR, standard error of the estimate (WSE), standard error of slope with user selected confidence interval (WCI) and R-squared value (WR2) are obtained [34]. The results of this method are controlled by the points with smaller positional uncertainty on the best-fit regression line [65]. If no values are provided by the user, DSAS uses the default uncertainty value. The weight (w) is defined as a function of the variance in the uncertainty of the measurement (e) [34]:

$$w = 1/\left(e^2\right) \tag{4}$$

Where e = shoreline uncertainty value.

Coefficient of Determination (R^2): It is the percentage of variance in the data that is explained by a regression [65]. It is used to verify the quality of the best-fit line regression.

3.2.3 Calculation of data uncertainty

The errors or uncertainties that arise due to different data sources, time of data acquisition, and the type of shoreline indicator were quantified based on several studies [73, 75]. According to Fletcher et al. [75] and Romine and Fletcher [76] there are two types of uncertainty: positional (seasonal and tidal fluctuations) and measurement (digitizing, pixel and rectification error). The uncertainty for each dataset was worked out considering the data product with due weightage of the quality of each data. The total uncertainty is used to calculate the weight and further working in the DSAS. Different uncertainties are explained below.

Seasonal error (Es): It is the error that arises due to seasonal changes of shoreline positions under the action of the waves and storms [75]. In the present study, all the scenes are of the winter season and hence this error was neglected.

Tidal fluctuation error (Et): It is the error associated with horizontal variability in shoreline position due to tides [75]. All the images in the present study correspond to high-tide values. Based upon the values the tidal range was considered as 2.12 m.

Digitizing error (Ed): It is the error related to shoreline digitization [75]. The digitizing error was kept within half a pixel of Landsat data (15 m).

Pixel error (Ep): It relates to image precision (resolution). In the present study except for the Landsat MSS data all the TM, ETM+ and OLI data have a spatial resolution of 30 m. To make the pixel uniform the 60 m MSS pixels were resampled to 30 m. Thus, the average pixel error was neglected.

Rectification error (Er): It is the square root of the mean error of the image rectification process [75, 76]. The rectification error in the level-2 Landsat products was found to be one fourth of a pixel, i.e., in the present case was decided as 7.5 m.

Total Positional Uncertainty: The total positional uncertainty (U_t) is the result of all errors that were previously estimated. It is defined as the square root of the sum of the squares of the sources of different errors [75, 76]. The formula of U_t is given as follows:

$$U_t = \pm\sqrt{\left(E_s^2 + E_t^2 + E_d^2 + E_p^2 + E_r^2\right)} \tag{5}$$

Where E_s is the seasonal error, E_t = tidal error, E_d = digitizing error, E_p = pixel error, and E_r = rectification error. The annualized uncertainty (U_a) was calculated using the square root of the sum of the squares of total positional uncertainty for each shoreline divided by the analysis period [75] as is given below.

$$U_a = \pm\frac{\sqrt{\sum_1^n U_{ti}^2}}{T} \tag{6}$$

Uncertainty	Positional uncertainty		Measurement uncertainty			Total positional uncertainty
	E_s (m)	E_t (m)	E_d (m)	E_p (m)	E_r (m)	U_t (m)
Landsat images	0	±2.12	±10	0	±5	11.37
SOI topo map	0	0	±15	0	±15	21.21
Army topo map	0	0	±15	±15	±30	33.54

Table 3.
Uncertainties associated with shorelines obtained from different sources.

Various uncertainties in the historical shoreline position between 1948 and 2021 is given in **Table 3**.

The weight (w) is defined as a function of the variance in the uncertainty of the measurement (e). Weighted Linear Regression Rate (WLR) was computed using the total positional uncertainty values.

3.2.4 DSAS configuration

It consists of four main steps as is given below.

i. *Baseline definition*: An offshore baseline, at 200 m distance, almost parallel to the shoreline, was created in shape file format (.shp) with required attributes. The baseline is required to calculate the distance from shoreline to it at each orthogonal transect. The movement of all shorelines over the 48-year period (1973–2021) has been computed about the baseline. Due to the large uncertainty of the shoreline generated from Army topo maps, it was not considered for DSAS but used for future reference. When the shoreline moves landward to the baseline it is considered as erosion (denoted as negative values). In contrast, when the shoreline moves seaward, it is considered as accretion (denoted as positive values).

ii. *Collection of shorelines*: Each shoreline vector represents a specific position in time and space, hence each of them is assigned a date (in dd-mmm-yy format) in the shoreline feature-class attribute table.

iii. *Generation of transects*: Orthogonal transects were generated at 100 m intervals alongshore by using DSAS. At each of the representative locations, shore normal transects were generated against which relative changes in shoreline position were determined. A total of 1924 effective orthogonal transects were casted along the baseline from the south-west corner of the west bank to the south-east corner of the east bank in a clockwise direction and numbered (as Transects ID). The measurement transects that are casted by DSAS from the baseline, intersect the shoreline vectors. The points of intersection stores location and time information which are subsequently used to calculate the rate of change. The distances from the baseline to each intersection point along a transect are used to compute the distance and rate statistics. Some of the transects were removed from the analyses as they were falling on the confluence of the tributaries of the Hooghly River.

iv. *Calculation of change in shoreline position and other statistics*: Two types of statistics are generated by DSAS *viz*. distance (NSM) and rate (EPR, LRR, and WLR). The regional change rate is calculated by averaging the rates of changes from all the transects. The average coefficient of determination (R^2) and uncertainties of the annual rate-of-change (m $yr.^{-1}$) are computed at a 95% confidence interval (LCI95 or WCI95).

3.2.5 Prediction of tide condition

The date, time and height of tide were calculated using WXTide32 package. The height of tide is governed by the following harmonic equation given in the Manual of Harmonic Analysis and Prediction of Tides, special publication no. 98, US Department of commerce [77].

$$h = H_0 + \sum_{n}^{N} f_n H_n \cos\left[a_n t + (V_0 + u)_n - K_{n'}\right] \tag{7}$$

Where, h is the height of tide at any time t.
H_0 = the mean height of water level above datum used for prediction.
H_n = the mean amplitude of any constituent A_n.
f_n = the factor for reducing mean amplitude to year of prediction.
a_n = the hourly speed of constituent A_n.
t = the time, in hours, reckoned from beginning of year or prediction.
$(V_0 + u)_n$ = the Greenwich equilibrium argument of constituent A_n when t = 0.
$K_{n'}$ = the modified epoch of constituent A_n.
N = the number or constituents used for the particular station.
In this equation except *h* and *t*, all other parameters are considered as the harmonic constant for any particular year and the place. Using these Harmonic constant, the successive value of tide height can be generated at any point of time. WXTide32 data pertaining to Diamond Harbor was considered to be representative of the present study. **Table 2** provides the satellite overpass time versus the low and high tide timings.

3.2.6 Cluster analysis

Cluster analysis is a technique used to classify cases into groups that are relatively homogeneous within themselves and heterogeneous between each other, based on a defined set of variables [78, 79]. Hierarchical agglomerative clustering using the Ward linkage method was followed in the present study. In this method, clusters are merged to reduce the variability within the cluster. At every stage the average similarity of the cluster is measured. A case is selected to enter the cluster if the inclusion in the cluster produces the least increase in the error. The number of the cluster centres was determined from 'Scree diagram' in which 'distance coefficients' are plotted against the 'stages'. The point at which there is a significant jump in the distance values was considered as the 'elbow' of the 'Scree plot'. The numbers of clusters were decided as the number of cases minus the step of the elbow. Once the clustering is done, K-mean classification is performed for all the transects using the number of cluster centres from 'Scree plot'. K-mean classification assign cluster membership and distance from the cluster centre to each case. Distance of the cluster centres are determined by using Euclidean distance as is given below:

$$d_{ij} = \left[\sum_{l=1}^{q} \left(x_{il} - x_{jl} \right)^2 \right]^{\frac{1}{2}} \tag{8}$$

Where dij = ED for two individuals i and j, each measured on q variables, Xij, Xji, $i = 1, \dots q$. ED (dij) is calculated as the sum of squared differences between relative cross-shore positions at each transect (T_1, T_2, T_3, etc.) during each epoch (a) and for all epochs (N). Smaller ED values indicate the cases are more similar. To evaluate the robustness of the clusters Kruskal-Wallis one-way ANOVA test [80] was carried out. A detailed description of cluster analysis can be found in Everitt et al. [81] and Hennig et al. [78].

4. Results and discussion

4.1 Shoreline configuration

The 200 km stretch of the study region has varied beach types including wide sandy beaches to mudflat, the mixture of sand and mud, mangrove wetlands as well as open mixed jungle at the backdrop of sandy/muddy beaches. The considerable length of the shorelines has embankments (**Table 4**). The western bank consists mainly of sandy and muddy beaches whereas the east bank predominantly consists of a muddy and mangrove systems with intermittent gap areas where the beach is absent. Zone-wise brief description of the beach configuration is given below.

Zone-1: This zone constitutes the western bank of Hooghly estuary and falls between the outlet of Pichhaboni Khal and Rashulpur river, confined between the transects T_{25} and T_{211}. The foreshore is mostly sandy. Several sandy beaches (near Haripur, Boral, Bankiput and Digene) and the large number of aquaculture ponds are present in this zone. A large difference between high and low water lines (maximum of about 850 m) is observed between Pichhaboni Khal to the south of Gopalpur based on Survey of India topomaps. From satellite imagery, significant accretion and expansion of forest land can be seen between the Pichhaboni outlet and the north-east of Junput.

Zone-2: This zone is represented by the transect number T_{218} to T_{524} and falls between the confluence of the Rashulpur river and Haldi river on the west bank of Hooghly. The foreshore is mostly muddy with small sandy beaches near Hijli and Khejuri. Thick forest vegetation is observed in the east of Thanaberia along the coast. Large numbers of aquaculture ponds are present between Kandlamari and Khejuri. There is no embankment present in the northern half of this zone. The difference

Statistics	1973–1980	1980–1989	1989–1995	1995–2000	2000–2005	2005–2010	2010–2017	2017–2021
Mean	−14.32	−15.43	1.69	−43.00	39.79	−7.64	42.06	70.67
sd	200.03	86.59	77.90	110.99	122.09	100.35	135.17	162.57
Max	1386.98	566.43	681.41	629.64	1141.27	908.46	1000.23	1057.30
Min	−404.98	−527.71	−293.49	−1061.84	−313.23	−478.47	−519.50	−180.76

Table 4.
Mean shoreline change (m) over different time intervals.

between high and low water lines is comparatively less (~ 80 m) between Talpati Khal and Haldi river.

Zone-3: This zone extends from the Haldi river to the Rupnarayan river confluence, demarcated by the transects T_{534} to T_{917}. This zone is highly convex towards the bay and the foreshore is muddy. The difference between high and low water line increases from Gilabaria (T_{600}) till Jhikarkhali (T_{685}) which further increases near Horkhali. Series of brick kilns and few aquaculture ponds can be seen in this zone. There is no protective embankment present is this zone.

Zone-4: This zone falls in the east bank of Hooghly estuary between the transect T_{919} and T_{1174}. Very narrow muddy foreshore in present here, however, in some places it is absent. Most of the river bank of this zone is embanked especially upto Kantabaria, however, from here till Kulpikata Khal there is no embankment. The narrow difference exists between high and low water lines except near Diamond Harbor. Series of brick kilns can be observed in the north (Simulbaria to Diamond Harbor) and south (Kantabaria to Kulpi) of this zone.

Zone-5: The zone covers the coastline between Kulpi and Kakdwip, defined by the transects T_{1175}-T_{1454}. The difference between high and low water lines is very small and sometimes absent. Almost the entire coastline has an embankment. Few brick kilns and aquaculture ponds are present in this zone.

Zone-6: This zone falls between Kakwip and Namkhana (T_{1455}-T_{1574}), totally embanked and shortest among all. The difference between high and low water lines is very small especially at south of Kakdwip and near Nadabhanga Khal. No brick kiln or aquaculture pond is present here.

Zone-7: This is the last and longest zone between Namkhana and Henry Island in the east bank of Hooghly, represented by the transect T_{1575}-T_{1967}. The foreshore consists mostly of a muddy area (Dakshin Durgapur) but a narrow sandy beach is present in the west of Namkhana and a wide sandy beach in the extreme south near Lakshmipur till the outlet of the Bakkhali river. In the south of Patibunia considerable area is under a mangrove swamp and open mixed jungle. Moderate to good forest exists in the extreme south of the region confined between T_{1912}-T_{1967}. Near Dakshin Durgapur, the difference between high and low water line is more (~300 m). Most of the coastline is embanked. No brick kilns and aquaculture ponds are present in this zone.

4.2 Spatiotemporal change in shoreline

The large difference in the shoreline position was observed within each time interval and among different intervals. The dynamics of the shoreline are mainly due to disequilibrium in the morphological state and northward tapering nature of the estuary coupled with plausible subsidence due to auto-compaction of the Holocene sediments. One commonality among all the time intervals is the large variation in the seaward end of both the banks (**Figure 3**). During 1995–2000 and 2005–2010, the overall variation in the shoreline position is minimum. In comparison to the east bank west bank has more variation except for 1973–1980. Considering all the temporal intervals between 1973 and 2021 average recession is maximum in 1995–2000 ($-$ 43 m $\pm$ 110.99) especially due to erosion in the southern part of the east bank of the estuary. In contrast, there is an increasing trend in the seaward extension of the shoreline since 2010. Between 2017 and 2021 the average accretion is 70.67 m ($\pm$ 162.57). The maximum accretion length was 1386 m at T_{1486} (south-west of Kalinagar) in 1973–1980 whereas maximum erosion was $-$1062 m at T_{1865} (west of Fraserganj) during 1995–2000 (**Table 4**).

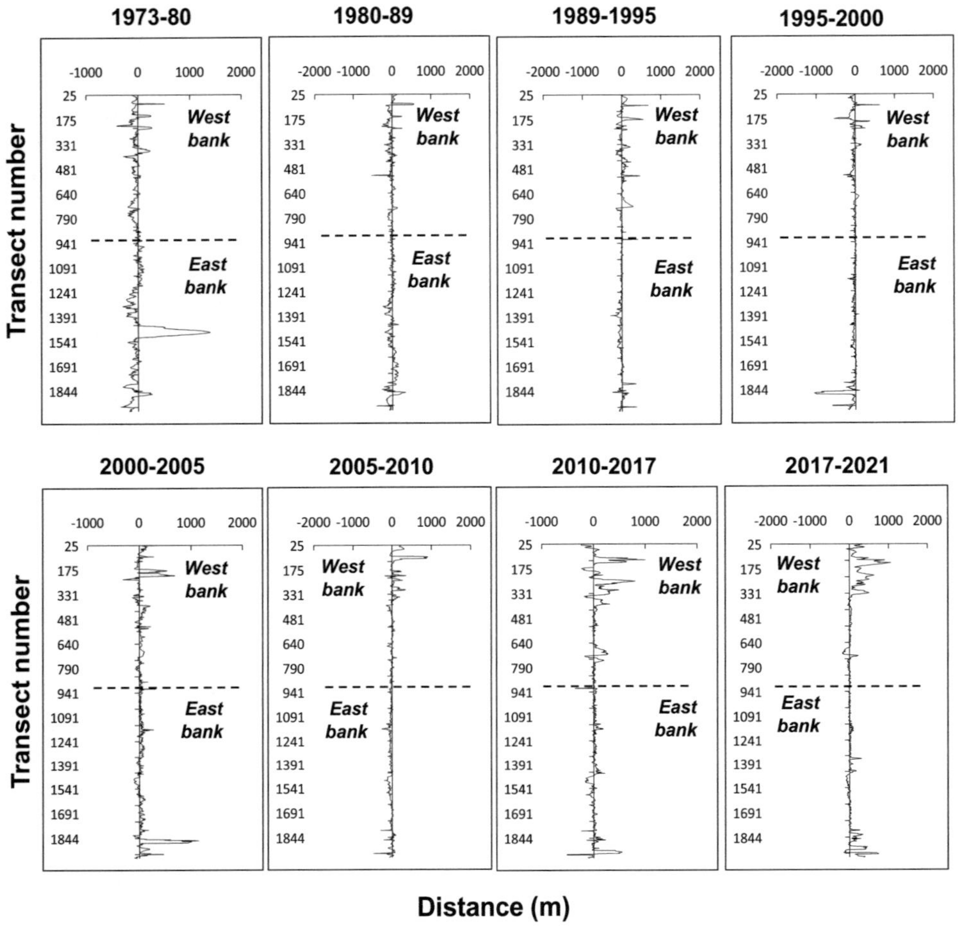

Figure 3.
Shoreline changes recorded at different transects over different temporal intervals.

The percentages of transects recorded aggradation or recession is given in **Figure 4**. From the figure, it is apparent that the proportion of aggradation and erosion does not match over the time intervals. The percentage of the transects exhibiting erosion was comparable during 2000–2005 (29%), 2010–2017 (30.20%) and 2017–2021 (27.81%). There was an abrupt increase in the erosion by 69.91% in 2010–2017. In general, there is a decreasing trend of erosion, especially after 2000 (**Figure 4**).

Figure 5 depicts how each zone contributes to the total shoreline change. Between 1973 and 2021, zone 5 contributed maximum towards erosion. Other zones that contributed marginally to erosion include zone 7 and zone 6. Zone 6 showed consistent erosion in all the intervals except for 1973–1980. Very high annualized aggradation of 69.17 m and 29.93 m was recorded in zone 1 and 2 respectively over the entire period of 1973–2021.

It is interesting to note that while comparing the coastline of 2021 with respect to 1948 (not used in the DSAS), there is a significant recession (~900 m) in the zone 2 (between Talpati Khal and Kaldalmari) and in the zone 3 (near Horkhali) by about 600 m. In the east bank, most significant erosion is noticeable in zone 5,

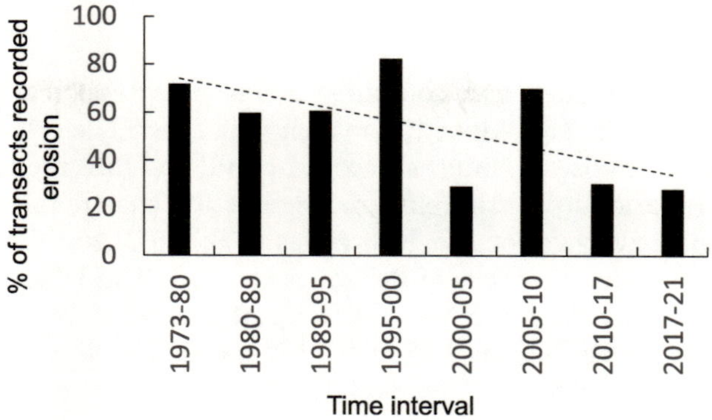

Figure 4.
Percentage of transcets showing erosion at different time intervals.

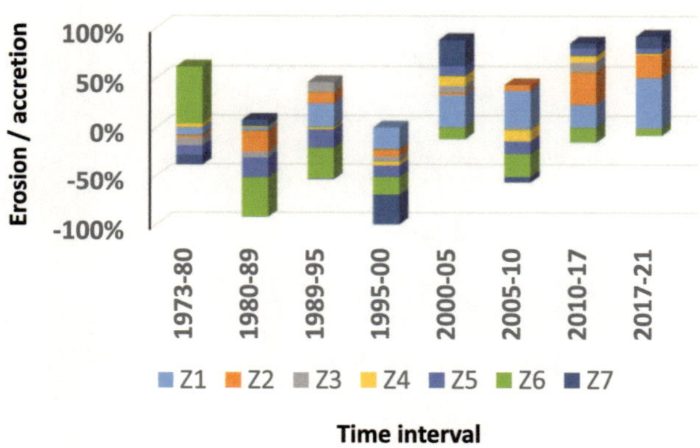

Figure 5.
Contribution of each zone towards erosion / accretion at different time intervals (Z represents the zones).

between Jadabnagar and Tilakmandal chak. The maximum landward retreat recorded was 2700 m near Uttar Chandannagar. On the other hand, accretion was observed in the south of zone 1 and 2 as well as in the north of zone 6. Quantitative analysis of the coastline change in this region has been carried out by Bandyopadhyay et al. [82], Raju et al. [83], Jana et al. [84], Rudra [85], Chakraborty [49] and Das et al. [86] along with their underlying mechanism. They have opined that beach erosion is attributed to various causes such as decrease of sediment supply from rivers, land subsidence, and interruption of longshore sediment transport by man-made structures. As the sea level rises, it causes waves to act on higher parts of the beach profile, resulting in enhanced erosion. If the sandy beaches disappear as a result sea-level rise, waves and storm surges, it will impact higher areas along the coastline [87].

Jana and Bhattacharya [88] used multi-resolution Landsat satellite imagery of 1972–2010 for shoreline change study along the 65 km long coastal stretch located between Rashulpur (Purba Medinipur) and Subarnarekha (Balasore) estuarine complex. The authors revealed that about 23 km of coastline recorded accretion, which was observed on several beaches such as at Talsari, Udaipur and Haripur, which were not affected by anthropogenic activities.

4.3 Shoreline change rate

The shoreline change rates were computed by linear regression and end point rate method at a lateral spatial interval of 100 m along the coast. The rates of changes of shoreline at different transect points estimated by EPR and LRR methods are given in **Figure 6**. Large variation in net shoreline movement and change rates were observed in the study region among various analysis zones (**Table 5**). Considering long term change between 1973 and 2021 four of the zones *viz.*, 1, 2, 3 and 4 showed positive change (aggradation) by WRR method, the range of which varies between 0.24 m $yr.^{-1}$ (zone 3) to as high as 9.45 m $yr.^{-1}$ (zone 1). The very high recession was found in east bank at zone 6 (− 4.35 m $yr.^{-1}$), followed by zone 5 (−3.02 m $yr.^{-1}$).

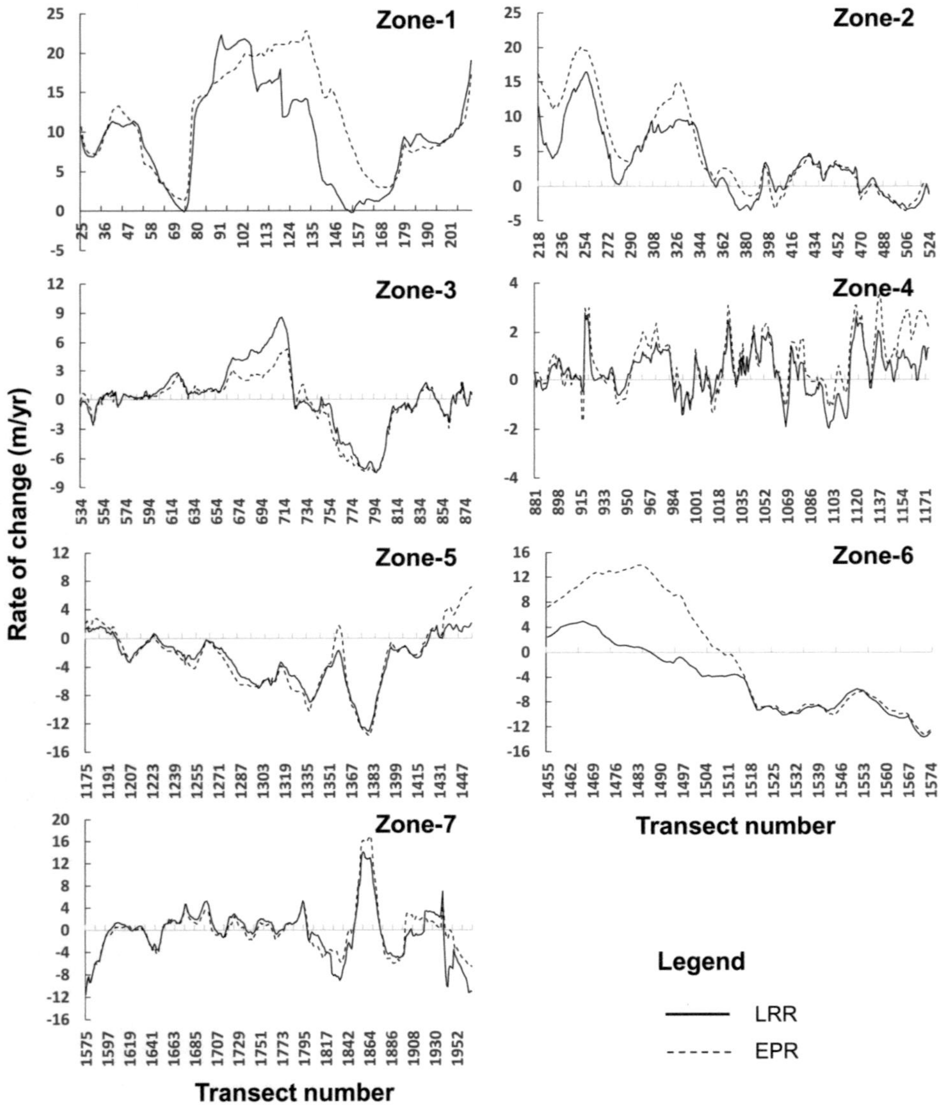

Figure 6.
The rate of change of shoreline by WRR and EPR method.

Zones	SCE	NSM	EPR	WLR
Zone-1	740.83 ± 359.22	553.34 ± 298.84	11.48 ± 6.20	9.45 ± 6.22
Zone-2	391.0 ± 263.97	239.42 ± 307.19	4.97 ± 6.38	3.47 ± 4.89
Zone-3	170.61 ± 128.57	−16.81 ± 137.25	−0.35 ± 2.85	0.24 ± 3.28
Zone-4	81.31 ± 41.47	36.91 ± 54.96	0.77 ± 1.14	0.36 ± 0.93
Zone-5	249.19 ± 150.22	−137.22 ± 206.24	−2.85 ± 4.28	−3.02 ± 3.54
Zone-6	668.09 ± 351.75	−5.18 ± 456.16	−0.11 ± 9.47	−4.35 ± 5.36
Zone-7	303.21 ± 259.99	−2.30 ± 210.13	−0.05 ± 4.36	−0.38 ± 4.47

Table 5.
Zone-wise average shoreline change envelope (SCE), net shoreline movement and change rate by EPR and WRR method.

Overall, the rate of aggradation superseded the rate of erosion in the 48 years span. The zones which experienced maximum net seaward movement include zone 1 (553.34 m) whereas maximum net landward movement (erosion) of shoreline was found in zone 5 (137.22 m) (**Table 5**). There is a good agreement between both the methods (EPR and WRR) in respect of zone 2, 3, 4, 5 and 7. In zone 1 large difference in the shoreline change rate calculated by both the methods was recorded between Junput and Jagannathpur (T_{102} to T_{168}), whereas in zone 6, the differences are significant in the region south of Kakdwip to Budhakhali along the Kakdwip river (T_{1455} to T_{1511}).

Although, the net shoreline movement (NSM) values are less in zones 3, 6 and 7 but the shoreline change envelope records large variation which indicates that the inter-annual fluctuation is very high in these zones and morphodynamic processes are very active.

Based upon the rate of erosion/accretion by WRR method, the transects were grouped into 7 classes (**Table 6**). From the table, it is evident that most of the shoreline (more than 73.33% by WRR and 69.95% by EPR) exhibit erosion/accretion rate between −5 and + 5 m $yr.^{-1}$. Low erosion rate (< 1.0 m/yr) was exhibited by 13.46% and 11.43% of the shoreline in WRR and EPR method respectively (not presented in the table). The proportion of very high erosion (<−10 m $yr.^{-1}$) and aggradation (>20 m $yr.^{-1}$) is limited to less than 2% of the shoreline. The spatial

Class	Range	WRR method		EPR method	
		No of transects	% of total transects	No of transects	% of total transects
1	< −10	37	1.92	28	1.45
2	−5 to −10	187	9.71	204	10.60
3	−5 to +5	1411	73.33	1346	69.95
4	5 to 10	166	8.62	121	6.28
5	10 to 15	77	4.00	130	6.75
6	15 to 20	29	1.50	71	3.69
7	> 20	17	0.88	24	1.24

Table 6.
Different classes of erosion/accretion rates and their contribution to the shoreline.

distribution of different change classes by WRR method is given in **Figure 7a**. It can be seen from the figure that in the west bank only one segment exhibits high erosion (−10 to −5 m yr.$^{-1}$) whereas in the east bank at least 6 segments (east of Kharibaria) show high erosion. This area exhibits has a large difference between low and high tide lines. While comparing with the Army Series map of 1948, it was found that there is a significant landward movement of shoreline between 1948 and 1973. In the east bank, there is no area under high erosion in zone 4, however, in zone 5, 6, and 7 considerable area along the shoreline is under high to very high erosion state. There are 3 distinct stretches near Uttar Chandannagar, Ramganunagar, Madhusudanpur and Lakshimipur. Close observation with the Army toposheet of 1948 reveals that there is an extensive recession in this area. The Rangatala island which used to be an integral part of the east bank has almost reduced to half between Kulpi and Madhusudanpur. The southern half of zone 6 has a high to very high rate of erosion between Budhakhali and north or Namkhana. The zone 7 is punctuated by two major areas of high erosion i) in the west of Edward creek, dominated by mangrove swamp and open mixed jungle and ii) in the east of Henrys island. In contrast to erosion, high to very high aggradation (> 20 m yr.$^{-1}$) is recorded between south of Gopalpur to Junput dominated by a wide sandy beach and inter-tidal difference. High aggradation is also observed in the south of the Rashulpur river confluence. In zone 2 high rates of accretion is observed in the north of Rashulpur river and east of Nij Kasba. In the east bank, there is no area of high accretion except in zone 7, near Lakshmipur dominated by mangrove swamps. This observation is in good agreement while comparing with the Army topo map of 1948.

4.4 Temporal pattern of erosion/accretion

To understand the temporal pattern of change direction, transects were grouped into two categories *viz*., eroding and aggrading type based upon displacement direction in each time interval. From 8 different temporal intervals 256 unique combinations were generated which were further grouped into 8 categories *viz*. i) consistent erosion (when in all the 8 temporal intervals the changes are negative) ii) mostly erosion (when erosion is recorded at least in 5-time intervals) iii) recent erosion (when last 3 or more consecutive intervals erosion is dominant) iv) mostly accretion (when accretion is recorded at least in 5-time intervals) v) recent accretion (when last 3 or more consecutive intervals accretion is dominant) vi) alternate (when erosion and accretion takes place alternatively) vii) trend reversal (changes from erosion to accretion over the years in a consecutive manner) and viii) others (when no definite trend is observed). **Table 7** provides change patterns and their contribution to the entire shoreline under study. At a long-temporal scale, 1.87% of the shoreline shows consistent erosion which is alarming and another 36.69% are mostly eroded. We could not find any transect recording accretion in all the time intervals. All the 3 accretion types (MA, RA and TREA) together account for 38.93% of the shorelines (**Table 7**). Reversal of trend towards aggradation during last 3 or more consecutive time intervals was found from 4.73% of the shoreline, mostly located in zone 1 and 2. Only a small proportion of the shoreline (0.42%) exhibits alternate erosion and accretion over the years and does not yield a definite trend. **Figure 7b** shows the changing pattern along the shorelines.

Some of the transects that recorded both high erosion rate (more than 5 m yr.$^{-1}$) and consistent erosion are located in the north of Sibkalinagar (T_{1372}-T_{1374}), south of

Type	Description	Change direction	No of transects	% of total shoreline
CE	Consistent erosion	-ve	36	1.87
ME	Mostly erosion	-ve	706	36.69
RE	Recent erosion	-ve	10	0.52
MA	Mostly accretion	+ve	571	29.68
RA	Recent accretion	+ve	87	4.52
ALT	Alternate	mixed	8	0.42
TREA	Trend reversal (erosion to accretion)	+ve	91	4.73
OTH	Others	mixed	415	21.57

Table 7.
Temporal change pattern of shoreline behavior and their contribution.

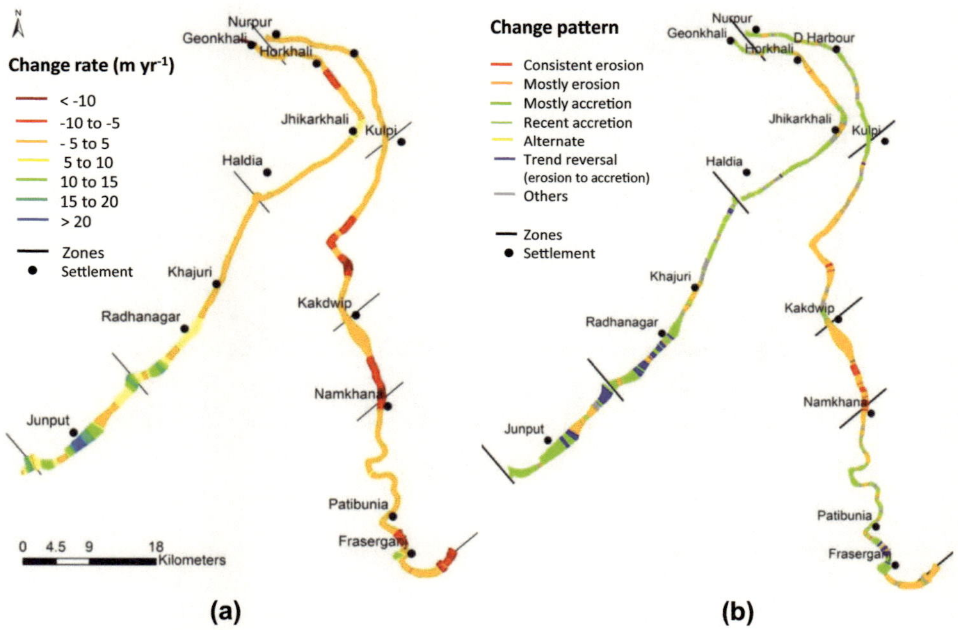

Figure 7.
Shoreline changes a) rate of erosion/accretion (m yr.$^{-1}$) and b) change pattern.

Budhakhali near Ghiya Khal (T_{1520}-T_{1533}), south of Nadabhanga Khal (T_{1552}-T_{1557}) and north of Duaragra Gang in zone 6 (T_{1569}-T_{1574}).

4.5 Hierarchical agglomerative clustering

Although, shoreline change analysis quantifies rates and directions of change, further analyses are needed to resolve distinct modes of coastal system behavior. Traditional shoreline changes analyses quantify the rate and direction of change by analyzing multi-date/historical data. However, there are some commonalities in terms of coastal system behavior. The Hierarchical agglomerative clustering was performed using the change matrix of all 8 temporal intervals to define the distinct coastal change behavior. Clustering was done using the Ward method which computes the sum of squared distance within the clusters and aggregates the clusters with the minimum increase in the overall sum of squares. The distance coefficients were plotted against the stage to generate a 'Scree diagram' (**Figure 8**). The number of clusters in the present study was 5 which was used for K-mean clustering. The cluster centres and the distances between cluster centres are given in **Tables 8** and **9** respectively.

The clusters captured a unique pattern of change at a temporal scale (**Table 8**). Among all the transects, 79.15% are represented by cluster 1 and only 0.94% by cluster 5. In clusters 1, 2 and 3 most of the transects show a balancing act of aggradation and erosion at different temporal intervals. The transects that recorded consistent erosion (**Figure** 7) were found in cluster 1 only. In cluster 4, erosion is dominant, while in cluster 5 accretion is dominant in most of the time span. The mean displacement of the

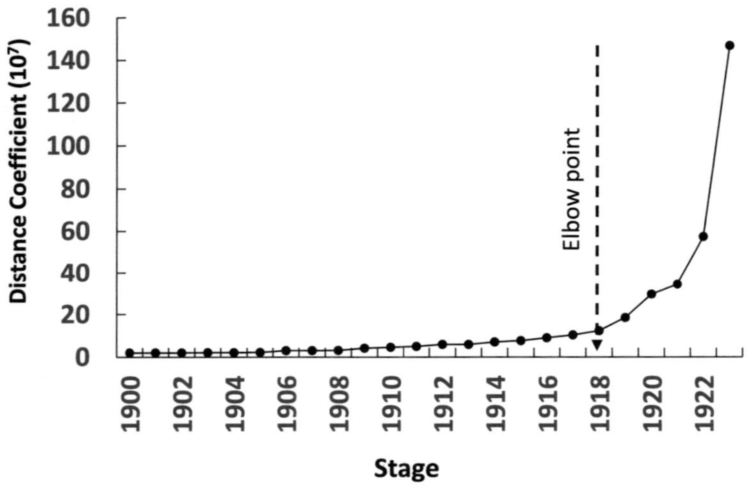

Figure 8.
Scree diagram defining the optimum number of clusters using elbow rule.

Temporal intervals	Clusters				
	1	2	3	4	5
1973–1980	−44.37	−52.95	−64.31	926.40	193.82
1980–1989	−6.50	−64.44	−51.74	−76.24	226.13
1989–1995	−.32	24.22	1.62	−57.19	82.35
1995–2000	−25.80	−89.62	−35.18	−50.58	−931.35
2000–2005	27.71	97.21	−14.32	−59.69	941.45
2005–2010	−21.90	85.34	48.19	−83.01	−3.82
2010–2017	32.63	−28.49	464.74	−87.87	22.23
2017–2021	12.53	341.60	445.03	−54.35	109.87

Table 8.
Various cluster centres.

Cluster	1	2	3	4	5
1	—	369	619	991	1335
2	369	—	521	1087	1278
3	619	521	—	1249	1476
4	991	1087	1249	—	1572
5	1335	1278	1476	1572	—
6	—	369	619	991	1335

Table 9.
Distance between cluster centres.

shoreline in cluster 1 is −3.25 m and the maximum is 99.25 m in cluster 3, constituting only 4.80% of the total transects. All the clusters show aggradation in terms of their mean displacement values except cluster 1.

5. Discussion

Beach profile morphology and coastline, change over a range of time and spatial scales. The short-term variability occurs over a period of days to a month as a result of i) episodic events (storms) ii) medium-term variability over several months (e.g., winter summer wave change) to several years (e.g., due to regional climate variability, engineering intervention and prevailing sedimentary processes) and iii) long term variability that occurs over a period of a decade to a century, associated mainly with climate change impact; and very long term millennial-scale evolution as a result of quaternary sea-level changes [89]. Broad-scale analysis of changes in shoreline position has the potential to highlight the role of regional forcing on large-scale coastal behavior, e.g., long-term tidal cycles [90] or sea level rise [4]. Shoreline change analysis is also useful to identify notable 'hotspots' of contrasting behavior [91, 92]. The Hooghly estuarine shoreline analyses studied here comprehend synthesis of historical shoreline change over 48 years supported by limited ground observations. The data has been analyzed at high spatial resolution (100 m, alongshore interval) along the entirety of a 200 km shoreline. In the area evidence for strong met-ocean forcing is ostensibly compelling. The phenomena of erosion and accretion are largely regulated by littoral current patterns and sediment influx from different rivers and the adjacent Bay of Bengal. The shoreline of this 200 km stretch has different configurations from the sandy beach to muddy swamp punctuated by anthropogenic footprints including brick kilns, aquaculture ponds, protective embankments and beach nourishment treatments. Beach nourishment projects and coastline protection structures can result in an artificial accretion of coastline in a short period [93]. Large variations exist in shoreline position within the same year and also among different years indicate the disequilibrium in the morphological state. There could be several external factors responsible for shoreline change including sea level rise, changes in the wave climatology and storm intensity as well as changes in the catchment characteristics due to deforestation and land degradation which results in higher sediment load in the terrestrial run-off. In contrast to surface runoff, engineering intervention through the construction of dams and barrages also makes the estuary sediment starved.

In long-term perspective, temporal data of PSMSL (Diamond Harbor and Haldia) reveals that the sea level is rising at the rate of 2.41 and 3.02 mm $yr.^{-1}$ respectively. The sea surface temperature induced El Niño-Southern Oscillation (ENSO) has a significant role in global atmospheric circulation influencing the temperature and precipitation. The irregular pattern of El Nino and La Nina triggers rainfall variability over the Indian sub-continent. In recent years strong La Nina and very strong El Nino have been witnessed in 2010–2011 and 2015–2015 respectively. The monsoon rainfall variability has a direct relation with terrestrial run-off and estuarine water level. Since 1951 there were 8 strong to very strong El Nino and 7 strong types of La Nina years. The storm surges are another strong forcing factor in a short temporal scale that can change the shoreline configuration. Although, the frequency of cyclonic storms is declining over the Bay of Bengal but the intensity is increasing. Extremely severe cyclonic storms of 2019 and 2020 are the best examples causing extensive damage to the coastline embankments. Karunarathna et al. [89] found single storms or storm clusters predominantly change the supra tidal and inter-tidal part of the beach profile and that beach erosion volumes are strongly correlated to the power of the storm. Once the astronomical tides coincide with storms, extreme sea level occurs resulting in large-scale inundation and damage to the coastal structures. Besides warming of sea surface relative, sea level change can also happen due to vertical land motion that can result from glacial isostatic adjustment, tectonic processes, coastal subsidence and uplift caused by anthropogenic factors. High-frequency and short temporal scale sea level variability due to seiches, meteotsunamis are frequently under-represented in sea level studies and yet contribute to the extreme sea levels which are of great research interest and importance to coastal dwellers [94]. In general, coastal landforms affected due to short-term perturbations *viz*. cyclone generally attains a morphodynamic equilibrium often by adopting different 'states' in response to varying wave energy and sediment supply [95]. Nevertheless, elevated surge water levels are known to be important drivers of longer-term dynamics on sedimentary shorelines [96]. Besides sea level rise and storms alongshore sediment transport can also have an impact on the coastline change, in particular, it is likely to result in coastline accretion.

Most of the west bank of Hooghly estuary is prograding at the rate of 0.24 m $yr.^{-1}$ in zone 3 to as high as 9.45 m $yr.^{-1}$ in zone 1. Whereas recession is pre-dominant in the east bank, especially in zone 5, 6 and 7 accounting −0.38 to −4.35 m $yr.^{-1}$. In general, aggradation dominates over erosion. Large variation in the shoreline change envelope in zone 3, 6 and 7 reveals an active morphodynamic process. The different suite of behaviors in recent intra-decadal scale suggests that forcing of coastal change can be interpreted as a form of the time-dependent complex response of the kind envisaged by Schumm and Lichty [97] whereby changes over shorter time scale, are inherently associated with tighter cause-effect linkages at smaller spatial scales, and broader trends emerge over longer time-scales. Additionally, the phenomena of erosion and accretion are largely regulated by littoral current patterns and sediment influx from different rivers and the adjacent Bay of Bengal. The west bank of the estuary having sandy inter-tidal plain is aggrading over longer time scale whereas several areas in the east bank of muddy beaches record the high rate of erosion. The temporal pattern of erosion/accretion has been captured using the direction of change in each time interval. Some portions of the shorelines especially north of Kakdwip and Namkhana recorded a consistent high rate of erosion (> −5 m $yr.^{-1}$) over each interval. Although, only 1.87% of the area of the shoreline showed consistent erosion for all the time intervals but together with 'mostly erosion' type it constitutes 38.56% which is alarming. These areas need to be protected from anthropogenic intervention and to be stabilized by rejuvenating protective embankments or vegetative barrier. Contrasting modes of prograding stretch adjacent to retreating stretch can be found in close proximity, particularly in zone 1 and 2, which suggests that local influences may be particularly important. Both these transitions in behavior suggest localized net littoral fluxes of sand and gravel from the north of estuary to the south-west. These localized instances of coupled behavior have led to a distinct net change in regional shoreline planform over a longer time scale. Some of the stretches of the shoreline exhibit distinct change of cuspate foreland from rounded to sharp apex especially north of Jhikarkhali and Madhusudanpur, north of Kakdwip. Erosion at the north and progradation at the west and south-west, illustrates south-west transport of sediments over the studied time scale whereas diffusive behavior dominated decadal-scale shoreline change.

The inter-temporal analysis using spatial smoothing windows of 1000 m showed that there is no consistent association between convexities/concavities and the erosion/accretion. Some concave stretches of shoreline exhibit erosional signatures, whilst others are accretional. The convexity of the shoreline near Horkhali (in the west bank) increased over time but decreased near Madhusudanpur on the east coast, however near Kakdwip and Patibunia the convexity remained almost unchanged over the years. Some of the concave stretches of the shoreline showed seaward accretion in the west bank, e.g., at Nij Kasba. The eroding sediments move parallel to the coast by alongshore currents from north to south direction and are expected to deposit around the concave coast owing to the lower current velocity [93]. As a result, the coastline can advance to the ocean around these regions. Several studies claim that a concave-shaped coastline tends to exhibit accretion while a convex-shaped coastline tends to exhibit erosion [93]. However, in the present study, several concave stretches of the east coast exhibited landward retreat of coastline typically along the Rangafala channel near Lakshmipur, between Ghiya Khal and Duraragra Gang (north of Namkhana) and small patches in Patibunia island. Presumably, both diffusive and anti-diffusive (unstable) behavior is operational [98] which are likely to change as the shoreline planform adjusts in response to the consequent patterns of erosion and deposition.

With the anticipated increase in global mean temperature by about 0.5°C, the thermal expansion and melting of ice caps and glaciers are inevitable [13] but this effect may be masked by inlet dynamics and coastal engineering projects even over extended time periods. However, the implication is that sea level rise is a secondary but inexorable cause of beach erosion in such areas which may lead to high-energy swells to reach further up the beach and redistribute sand offshore. Apart from the external and natural forces there are alarming uncontrolled anthropogenic activities which have imposed excessive pressure on the coastal landuse and exacerbating beach erosion problems along the Hooghly estuary. This will have ominous implications for ever-increasing coastal population and associated livelihood [99]. There is a need for decoupling the long-term forces from the anthropogenic effects and projecting the future scenario of coastal changes for effect coastal planning and enforcement.

6. Conclusion

The study of historical evolution and sculpturing of the coastal areas of Hooghly estuary in terms of short and longer time scale has significant importance in evaluating the criticality in shoreline change. The findings of the present study revealed that geospatial techniques are very useful for analyzing and predicting shoreline dynamics. The short- and long-term changes have been estimated using the DSAS extension tool of ArcGIS. The tool enables the calculation of several change metrics and also the rate of changes from time-series shoreline positions and helps in determining the zones of erosion and accretion. The variation is higher in the west bank than east bank except for 1973–1980. Considering the entire study period average recession is maximum in 1995–2000 (-43 m $\pm$ 110.99) especially due to erosion in the southern part of the east bank of the estuary. Zone 5 contributed maximum towards erosion, however, in general, there is a decreasing trend of erosion, especially after 2000. While comparing with 1948-topomaps there is a significant recession ($\sim$900 m) in zone 2 (between Talpati Khal and Kaldalmari) and in zone 3 (near Horkhali) by about 600 m. On the east bank, the most significant erosion is noticeable in zone 5, between Jadabnagar and Tilakmandal chak. The maximum landward retreat recorded was 2700 m near Uttar Chandannagar. The shoreline erosion is attributed to various causes such as decrease of sediment supply from rivers after construction of barrages in the upstream, land subsidence due to natural compaction or extraction of ground water, interruption of longshore sediment transport by man-made structures and dredging operation to maintain the navigation channel. In contrast, there is an increasing trend in the seaward extension of the shoreline since 2010. Between 2017 and 2021 the average accretion is 70.67 m ($\pm$162.57). Very high annualized aggradation of 69.17 m and 29.93 m was recorded in zone 1 and 2 respectively over the study period. The shoreline change rate computed using WLR method reveals that zone 1, 2, 3 and 4 show the positive change (aggradation) which varies between 0.24 m yr.$^{-1}$ (zone 3) to as high as 9.45 m yr.$^{-1}$ (zone 1). The very high recession was found in the east bank in zone 6 ($-$ 4.35 m yr.$^{-1}$), followed by zone 5 ($-$3.02 m yr.$^{-1}$). More than 73% area of the shoreline exhibits erosion/accretion between -5 and 5 m yr.$^{-1}$. The proportion of very high erosion (< -10 m yr.$^{-1}$) and aggradation (> 20 m yr.$^{-1}$) is limited to less than 2% of the shoreline. The temporal change pattern was examined using change direction in each time interval. About 1.87% of the shoreline shows consistent erosion in which at all the time-interval the direction of change was negative and an additional 36.69% constitutes of mostly eroded, characterized by erosion in at least 5 epochs. There is no area where consistent accretion was observed. In about 4.37% of the shoreline trend reversal from erosion to accretion has been observed. The change rate and pattern maps generated in the study will be helpful for policy makers to prepare a strategic coastal management plan and for future policy intervention. It is suggested that there should have a regular monitoring mechanism of this estuarine region to keep watch on the shoreline change and triggering factors and regulatory purpose.

Acknowledgements

The authors are thankful to the Chief General Manager, RRSCs (NRSC) for his keep interest and sustained support to carry out this study. Thanks, are also due to ea rthexplorer.usgs.gov for providing satellite data freely to the user community.

Author details

Dibyendu Dutta[1]*, Tanumi Kumar[1], Chiranjivi Jayaram[1] and Wasim Akram[2]

1 Regional Remote Sensing Centre – East (NRSC), Kolkata, India

2 Vidyasagar University, Midnapur, West Bengal, United States

*Address all correspondence to: ddisro@gmail.com

References

[1] Lockwood M. NSDI Shoreline Briefing to the the FGDC Coordination Group. Washington, DC: Project report to NOAA/NOS; 1997

[2] Boak, EH, Turner IL. Shoreline definition and detection: A review. Journal of Coastal Research. 2005;**21**(4): 688-703

[3] Richmond MD. A Guide to the Seashores of Eastern Africa and the Western Indian Ocean Islands. Stockholm, Sweden: Swedish International Development Cooperation Agency; 1997

[4] Zhang K, Douglas BC, Leatherman SP. Global warming and coastal erosion. Climate Change. 2004;**64**(1):41-58. DOI: 10.1023/B: CLIM.0000024690.32682.48

[5] Hanslow DJ. Beach erosion trend measurement: A comparison of trend indicators. Journal of Coastal Research. 2007;**50**:588-593

[6] Paterson S, Donnell AO, Loomis DK, Hom P. The Social and Economic Effects of Shoreline Change: North Atlantic, South Atlantic, Gulf of Mexico, and Great Lakes Regional Overview. Lexington, MA: Eastern Research Group, Inc; 2010. p. 02421

[7] Regnauld H, Pirazzoli PA, Morvan G, Ruz M. Impacts of storms and evolution of the coastline in western France. Marine Geology. 2004;**210**:325-337. DOI: 10.1016/j.margeo.2004.05.014

[8] Cooper JAG, Navas F. Natural bathymetric change as a control on century-scale shoreline behaviour. Geology. 2004;**32**:513-516

[9] Vassie JM, Woodworth PL, Holt MW. An example of North Atlantic deep-ocean swell impacting Ascension and St Helena islands in the central South Atlantic. Journal of Atmospheric and Oceanic Technology. 2004;**21**:1095-1103

[10] Galgano FA, Douglas BC, Leatherman SP. Trends and variability of shoreline position. Journal of Coastal Research. 1998;**26**:282-291

[11] Dronkers J. Dynamics of Coastal Systems. Advanced Series on Ocean Engineering. Vol. 25. Hackensack, NJ: World Scientific Publishing Company; 2005

[12] Williams SJ. Sea-level rise implications for coastal regions. Journal of Coastal Research. 2013;**63**:184-196. DOI: 10.2112/SI63-015.1

[13] Intergovernmental Panel on Climate Change (IPCC). Climate Changes: The Science of Climate Change, Summary for Policymakers and Technical Summary of the Working Group 1 Report. New York: Cambridge University Press; 1996

[14] Cendrero A. Mapping and evaluation of coastal areas for planning. Ocean and Shoreline Management. 1989;**12**:427-462. DOI: 10.1016/0951-8312(89)90023-4

[15] Ryan TW, Sementilli PJ, Yuen P, Hunt BR. Extraction of shoreline features by neural nets and image processing. Photogrammetric Engineering and Remote Sensing. 1991; **57**:947-955

[16] Lillesand TM, Kiefer RW, Chipman JW. Remote Sensing and Image Interpretation. Hoboken, N.J: John Wiley; 2015

[17] Braud DH, Feng W. Semi-automated construction of the Louisiana coastline

digital land/water boundary using Landsat Thematic Mapper satellite imagery. Department of Geography & Anthropology, Louisiana State University, Baton Rouge, Louisiana, USA. Louisiana Applied Oil Spill Research and Development Program, OSRAPD Technical Report Series, 97 002; 1998

[18] Frazier PS, Page KJ. Water body detection and delineation with Landsat TM data. Photogrammetric Engineering and Remote Sensing. 2000;**66**:1461-1467

[19] Liu H, Jezek KC. Automated extraction of coastline from satellite imagery by integrating canny edge detection and locally adaptive thresholding methods. International Journal of Remote Sensing. 2004;**25**: 937-958

[20] Di K, Ma R, Li R. Automatic shoreline extraction from high-resolution IKONOS satellite imagery. In: Proceedings of the ASPRS Annual Conference (May 2003). Anchorage: Alaska; 2003

[21] Chalabi A, Mohd-Lokman H, Mohd-Suffian I, Karamali K, Karthigeyan V, Masita M. Monitoring shoreline using IKONOS image and aerial photographs: A case study of Kuala Terengganu area, Malaysia. In: ISPRS Commission VII mid-term symposium "Remote Sensing: From Pixels to Processes". Enschede: The Netherlands; 2006. pp. 8-11

[22] Berberoglu S, Lloyd CD, Atkinson PM, Curran PJ. The integration of spectral and textural information using neural networks for land cover mapping in the Mediterranean. Computers & Geosciences. 2000;**26**:385-396

[23] Bastin L. Comparison of fuzzy c-means classification, linear mixture modeling and MLC probabilities as tools for unmixing coarse pixels. International Journal of Remote Sensing. 1997;**18**:3629-3648

[24] Zang J, Foody GM. A fuzzy classification of sub-urban land cover from remotely sensed imagery. International Journal of Remote Sensing. 1998;**19**:2721-2738

[25] Stuckens J, Coppin PR, Bauer ME. Integrating contextual information with per-pixel classification for improved land cover classification. Remote Sensing of Environment. 2000;**71**:282-296

[26] VLS. Feature Analyst Version 4.1 for Imagine. Reference Manual. Missoula, USA: Visual Learning Systems Inc.; 2007

[27] Harris PM, Ventura SJ. The integration of geographic data with remotely sensed imagery to improve classification in an urban area. Photogrammetric Engineering & Remote Sensing. 1995;**51**:993-998

[28] Stefanov W, Ramsey MS, Christensen PR. Monitoring urban land cover change: An expert system approach to land cover classification of semi-arid to arid urban centres. Remote Sensing of Environment. 2001;**77**: 173-185

[29] Vogelmann JE, Sohl T, Howard SM. Regional characterization of land cover using multiple sources of data. Photogrammetric Engineering and Remote Sensing. 1998;**64**:45-57

[30] Addo KA, Jayson-Quashigah PN, Kufogbe KS. Quantitative analysis of shoreline change using medium resolution satellite imagery in Keta, Ghana. Marine Sciences. 2011;**1**:1-9. DOI: 10.5923/j.ms.20110101.01

[31] Maiti S, Bhattacharya AK. Shoreline change analysis and its application to

prediction: A remote sensing and statistics-based approach. Marine Geology. 2009;**257**:11-23

[32] Nassar K, Mahmod WE, Fath H, Masria A, Nadaoka K, Negm A. Shoreline change detection using DSAStechnique: Case of North Sinai coast, Egypt. Marine Georesources & Geotechnology. 2019;**37**:81-95. DOI: 10.1080/1064119X.2018.1448912

[33] Nayak S. Use of satellite data in coastal mapping. Indian Cartographer. 2002;**22**:1

[34] Thieler ER et al. The Digital Shoreline Analysis System (DSAS) Version 4.0 an ArcGIS Extension for Calculating Shoreline Change. Reston, VA: U.S. Geological Survey; 2009

[35] Yu K, Hu C, Muller-Karger FE, Lu D, Soto I. Shoreline changes in west-Central Florida between 1987 and 2008 from Landsat observations. International Journal of Remote Sensing. 2011;**32**(23): 8299-8313

[36] Chen WW, Chang HK. Estimation of shoreline position and change from satellite images considering tidal variation. Estuarine and Coastal Shelf Science. 2009;**84**:54-60. DOI: 10.1016/j.ecss.2009.06.002

[37] Alesheikh A, Ghorbanali A, Nouri N. Coastline change detection using remote sensing. International Journal of Environmental Science & Technology. 2007;**4**:61-66. DOI: 10.1007/BF03325962

[38] Chandrasekar N, Viviek VJ, Saravanan S. Coastal vulnerability and shoreline changes for southern tip of India-remote sensing GIS approach. Journal of Earth Science and Climate Change. 2013;**04**:10. DOI: 10.4172/2157-7617.1000144

[39] Kumar A, Allu NC, Jayappa KS. Shoreline changes and morphology of spits along southern Karnataka, west coast of India: A remote sensing and statistics-based approach. Geomorphology. 2010;**120**:133-152. DOI: 10.1016/j.geomorph.2010.02.023

[40] Digital Shoreline Analysis System (DSAS). ver. 5.0 ver. 5.0 User Guide, Open-File Report 2018-1179, US Department of the Interior, US Geological Survey, 2018, 110 p

[41] Brooks SM, Spencer T. Temporal and spatial variations in recession rates and sediment release from soft rock cliffs, Suffolk coast, UK. Geomorphology. 2010;**124**:26-41

[42] Burningham H, French J. Understanding coastal change using shoreline trend analysis supported by cluster-based segmentation. Geomorphology. 2017;**282**:131-149

[43] González-Villanueva R, Costas S, Pérez-Arlucea M, Jerez S, Trigo RM. Impact of atmospheric circulation patterns on coastal dune dynamics, NW Spain. Geomorphology. 2013;**185**:96-109. DOI: 10.1016/j.geomorph.2012.12.019

[44] Sheik M. A shoreline change analysis along the coast between Kanyakumari and Tuticorin, India, using digital shoreline analysis system. Geo-spatial Information Science. 2011;**14**(4):282-293

[45] Patra A, Bhaskaran PK. Trends in wind-wave climate over the head bay of Bengal region. International Journal of Climatology. 2016;**36**:4222-4240. DOI: 10.1002/joc.4627

[46] Ghosh A, Mukhopadhyay S. Quantitative study on shoreline changes and erosion hazard assessment: Case study in Muriganga–Saptamukhi interfluve, Sundarban, India. Modelling

Earth Systems and Environment. 2016;**2**: 1-14. DOI: 10.1007/s40808-016-0130-x

[47] Bandyopadhyay S, Nandy S. Trends of sea level rise in Hugli estuary, India. Indian Journal of Geo-marine Science. 2011;**40**:802-812

[48] Biswas AN. Geohydro-morphometry of Hooghly estuary. Journal of Institute of Engineers (India). 1985;**66**:61-73

[49] Chakraborty S. Delineation of morpho-structural changes of some selected islands in the ganga delta region, West Bengal, India: A spatio-temporal change detection analysis using GIS and remote sensing. International Journal of Science and Nature. 2013;**4**:499-507

[50] Islam MR, Begum SF, Yamaguchi Y, Ogawa K. The Ganges and Brahmaputra rivers in Bangladesh: Basin denudation and sedimentation. Hydrological Processes. 1999;**13**:2907-2923

[51] Stanley DJ, Hait AK. Holocene depositional patterns, neotectonics and Sundarban mangroves in the Western Ganges–Brahmaputra delta. Journal of Coastal Research. 2000;**16**(1):26-39

[52] Goodbred SL Jr, Kuehl SA, Steckler MS, Sarker MH. Controls on facies distribution and stratigraphic preservation in the Ganges–Brahmaputra delta sequence. Sedimentary Geology. 2003;**155**:301-316

[53] Alam M, Alam MM, Curray JR, Chowdhury MLR, Gani MR. An overview of the sedimentary geology of the Bengal Basin in relation to the regional tectonic framework and basin-fill history. Sedimentary Geology. 2003; **155**:179-208

[54] Islam SN, Gnauck A. Mangrove wetland ecosystems in Ganges–Brahmaputra delta in Bangladesh. Frontiers in Earth Sciences in China. 2008;**2**:439-448. DOI: 10.1007/s11707-008-0049-2

[55] Chakrabarti P. Evolutionary history of the coastal quaternaries of the Bengal plain, India. Proceedings of Indian National Science Academy. 1995;**61**: 343-354

[56] Banerjee M, Sen PK. Palaeobiology in understanding the change of sea level and coast line in Bengal basin during Holocene period. Indian Journal of Earth Science. 1987;**14**:307-320

[57] Niyogi D. Geological background of beach erosion at Digha, West Bengal. Bulletin of Geology Mining and Metallurgical Society of India. 1970; **43**:1-36

[58] Gaur AS, Vora KH. Ancient shorelines of Gujarat, India, during the Indus civilization (Late mid-Holocene): A study based on archaeological evidences. Current Science. 1999;**77**: 180-185

[59] Mcfeeters SK. The use of the normalize difference water index (NDWI) in the delineation of open water features. International Journal of Remote Sensing. 1996;**17**:1425-1432. DOI: 10.1080/0143/1169608948714

[60] Xu H. Modification of normalised difference water index (NDWI) to enhance open water features in remotely sensed imagery. International Journal of Remote Sensing. 2006;**27**(14):3025-3033. DOI: 10.1080/01431160600589179

[61] Li W, Du Z, Ling F, Zhou D, Wang H, Gui Y, et al. A comparison of land surface water mapping using the normalized difference water index from TM, ETM plus and ALI. Remote Sensing. 2013;**5**:5530-5549. DOI: 10.3390/rs5115530

[62] Woodshole. (1990) Available from: http://woodshole.er.usgs.gov/project-pages/dsas/

[63] Thieler ER, Danforth WW. Historical shoreline mapping (II) application of the digital shoreline mapping and analysis systems (DSMS/DSAS) to shoreline change mapping in Puerto Rico. Journal of Coastal Research. 1994;**10**(3):600-620

[64] Theiler ER, Dadforth WW. Historical shoreline mapping (I): Improving techniques and reducing positioning errors. Journal of Coastal Research. 1994;**10**(3):549-563

[65] Himmelstoss EA. DSAS 4.0 installation instructions and user guide. In: Thieler ER, Himmelstoss EA, Zichichi JL, Ergul A, editors. 2009 Digital Shoreline Analysis System (DSAS) Version 4.0 — An ArcGIS Extension for Calculating Shoreline Change. Reston, Virginia, USA: U.S. Geological Survey Open-File Report 2008–1278; 2009

[66] Oyedotun TDT. Shoreline geometry: DSAS as a tool for historical trend analysis. In: Geomorphological Techniques. Vol. 3. London, UK: British Society for Geomorphology; 2014. pp. 1-12

[67] Chand P, Acharya P. Shoreline change and sea level rise along coast of Bhitarkanika wildlife sanctuary, Orissa: An analytical approach of remote sensing and statistical techniques. International Journal of Geomatics and Geosciences. 2010;**1**:436-455

[68] Faye I, Giraudet E, Gourmelon F, Henaff A. Cartographienormalisée de l'évolution du trait de côte. Mappemonde. 2011;**104**:12

[69] Jamont MF. Etude des aléas naturels sur le «Sud Vendée et marais Poitevin» - Rapport de phase 2 -Caractérisation des aléas de référence. France: Vendée; 2014. p. 67

[70] Prukpitikul S, Buakaew V, Keshdet W, Kongprom A, Kaewpoo N. Shoreline change prediction model for coastal zone management in Thailand. Journal of Shipping and Ocean Engineering. 2012;**2**: 238-243

[71] Honeycutt MG, Crowell M, Douglas BC. Shoreline-position forecasting: Impact of storms, rate-calculation methodologies, and temporal scales. Journal of Coastal Research. 2001;**17**: 721-730

[72] Anders FJ, Byrnes MR. Accuracy of shoreline change rates as determined from maps and aerial photographs. Shore and Beach. 1991;**59**:17-26

[73] Crowell M, Leatherman SP, Buckley MK. Historical shoreline change: Error analysis and mapping accuracy. Journal of Coastal Research. 1991;**7**:839-852

[74] Moore LJ. Shoreline mapping techniques. Journal of Coastal Research. 2000;**16**:111-124

[75] Fletcher CH, Romine BM, Genz AS, Barbee MM, Dyer M, Anderson TR, et al. National Assessment of Shoreline Change: Historical Shoreline Change in the Hawaiian Islands. Reston, Virginia: US Department of the Interiors, US Geological Survey; 2012. p. 55

[76] Romine BM, Fletcher CH. Armouring on eroding coasts leads to beach narrowing and loss on Oahu, Hawaii. In: Cooper JAG, Pilkey OH, editors. Pitfalls of Shoreline Stabilization: Selected Case Studies. Coastal Research Library. Dordrecht, Netherlands: Springer Science and Business Media; 2012

[77] Schureman P. Manual of Harmonic Analysis and Prediction of Tides, Special Publication no. 98. Washington: US Department of commerce; 1994

[78] Hennig, C. (2015). Clustering strategy and method selection. Available from: https://arxiv.org/abs/1503.02059

[79] Legendre P, Legendre L. Numerical Ecology, in Developments in Environmental Modelling. 3rd ed. Vol. 24. Amsterdam, The Netherlands: Elsevier; 1998

[80] Kruskal WH, Wallis WA. Use of ranks in one-criterion variance analysis. Journal of the American Statistical Association. 1952;**47**:583-621. DOI: 10.2307/2280779

[81] Everitt BS, Landau S, Leese M, Stahl D. Cluster Analysis. 5th ed. Chichester, UK: Wiley; 2011. p. 330

[82] Bandyopadhyay S, Mukherjee D, Bag S, Pal DK, Rudra K. 20th century evolution of banks and islands of the Hugli estuary. In: Singh S, Sharma HS, De SK, editors. Geomorphology and Environment. Kolkata, India: ACP Publishers; 2004. pp. 235-263

[83] Raju DK, Santosh K, Chandrasekar J, Tiong-Sa T. Coastline change measurement and generating risk map for the coast using geographic information system. The International Archives of Photogrammetry, Remote Sensing and Spatial Information Science. 2010;**38**(part II):492-497

[84] Jana A, Sheena S, Biswas A. Morphological change study of Ghoramara Island, eastern India using multi temporal satellite data. Research Journal Recent Science. 2012;**1**:72-81

[85] Rudra K. The Atlas of the Changing River Courses in West Bengal. Kolkata: Sea Explorers' Institute (SEI); 2012

[86] Das S, Choudhury MR, Das S, Khan S. Monitoring shore line and inland changes by using multi-temporal satellite data and risk assessment: A case study of Ghoramara Island, West Bengal. International Journal of Geoscience and Technology. 2013;**1**:1-20

[87] Nimura C. Rock art and coastal change in bronze age Scandinavia. In: Encountering Imagery: Materialities, Perceptions, Relations. Dept. of Archaeology and Classical Studies, Stockholm University; 2013. pp. 117-132

[88] Jana A, Bhattacharya AK. Assessment of coastal erosion vulnerability around Midnapur-Balasore coast, eastern India using integrated remote sensing and GIS technique. Journal of the Indian Society of Remote Sensing. 2013;**41**:12. DOI: 10.1007/s12524-012-0251-2

[89] Karunarathna H, Horrillo-Caraballo J, Kuriyama Y, Mase H, Ranasinghe R, Reeve DE. Linkages between sediment composition, wave climate and beach profile variability at multiple timescales. Marine Geology. 2016;**381**:194-208

[90] Gratiot N, Anthony EJ, Gardel AA, Gaucherel C, Proisy C, Wells JT. Significant contribution of the 18.6-year tidal cycle to regional coastal changes. Nature Geoscience. 2008;**1**:169-172

[91] Hapke CJ, Lentz EE, Gayes PT, McCoy CA, Hehre RE, Schwab WC, et al. A review of sediment budget imbalances along Fire Island, New York: Can nearshore geologic framework and patterns of shoreline change explain the deficit? Journal of Coastal Research. 2010;**26**:510-522

[92] McNinch JE. Geologic control in the near-shore: Shore-oblique sandbars and shoreline erosional hotspots, Mid

Atlantic Bight, USA. Marine Geology. 2004;**211**:121-141

[93] Xu N. Detecting coastline change with all available Landsat data over 1986–2015: A case study for the state of Texas, USA. Atmosphere. 2018;**9**(3):107. DOI: 10.3390/atmos9030107

[94] Woodworth PL, Melet A, Marcos M, Ray RD, Wöppelmann G, Sasaki YN, et al. Forcing factors affecting sea level changes at the coast. Surveys in Geophysics. 2019;**40**:1351-1397

[95] Woodroffe CD. Coasts: Form, Process and Evolution. Cambridge, UK: Cambridge University Press; 2003. p. 623

[96] Chaverot S, Héquette A, Cohen O. Changes in storminess and shoreline evolution along the northern coast of France during the second half of the 20th century. Zeitschrift für Geomorphologie, Supplementary. 2008;**52**:1-20. DOI: 10.1127/0372-8854/2008/0052S3-0001

[97] Schumm SA, Lichty RW. Time, space, and causality in geomorphology. American Journal of Science. 1965;**263**: 110-119

[98] Ashton AD, Murray AB, Littlewood R. The response of spit shapes to wave-angle climates. In: Sixth International Symposium on Coastal Engineering and Science of Coastal Sediment Process. Reston, VA: American Society of Civil Engineers; 2007. DOI: 10.1061/40926 (239)27

[99] Bird ECF. Coastline Changes: A Global Review. Chichester: John Wiley and Sons; 1985. p. 219

Chapter 3

Assessment of North Sinai Shoreline Morphodynamics Using Geospatial Tools and DSAS Technique

Ali Masria, Karim Nassar and Mohamed Galal Eltarabily

Abstract

This study employs a digital shoreline analysis system (DSAS) to identify and evaluate historical changes in the coastline along the North Sinai coast of Egypt. Using multi-temporal satellite images, change detection is explored along coastline over 27 years (1989–2016). The annualized uncertainty of shoreline changes was calculated. Erosion and accretion patterns were automatically quantified via four statistical parameters in the DSAS model namely net shoreline movement (NSM), rate of -8.17 m year^{-1} was recorded at the west seaside of El-Tinah plain throughout the 27 years. This recession of the shoreline is attributed to the joint effect of the stormy climate of the western seaside and the sediments transport from the Nile Delta. shoreline has progressed west of El-Bardawil inlet towards El-Arish harbor, where wave-induced littoral transport is ceased by the construction of jetties. The shoreline at the downdrift side of the jetties to the east has adversely retreated where the subsequent beaches are reverted at rates of -4.5 and -2.9 m year^{-1}. Lastly, the EPR model was utilized for quantifying shoreline changes in the near future of years 2025, 2035, and 2050.

Keywords: satellite imagery, remote sensing technique, morphodynamic detection, DSAS model, North Sinai Coast

1. Introduction

Coastal zones are now experiencing increased natural and human disruptions, such as sea-level rise, coastal erosion, and resource overexploitation, to name a few. Coastal erosion affects almost 80% of the world's beaches, with rates ranging from 1.0 cm year^{-1} to 30 m year^{-1}, posing a major threat to several coastal regions [1].

According to [2], increased knowledge of many driving forces is affecting the health of global coastal ecosystems has expedited efforts to evaluate, monitor, and reduce coastal stressors to understand the spatial distribution of erosion risks, predict their growth tendency, and support mechanism research on erosion and its solutions.

Shoreline extraction and change detection rates at different times are critical for coastal zone monitoring. The coastline, defined by [3] as the position of the land-water interface at a single point in time, is a highly dynamic characteristic that serves as a predictor of coastal erosion and accretion. Shoreline changes occur on a variety of time scales, ranging from geological to short-term catastrophic events. Waves, winds, tides, sea-level rise, frequent storms, geomorphic processes of erosion and accretion, and human activities are all factors that affect these changes [4].

Several international studies looked at quantitative and qualitative analysis of shoreline spatiotemporal fluctuations [5–13].

Alternatively, efforts were made to estimate the potential position of the shoreline to reduce the impact of the upcoming erosion activity. Moreover, for future predictions of shoreline spatial change, extensive and reliable information regarding historical and present coastline position is required. In a GIS framework, shoreline prediction models are simple to implement. With the help of historical data, several statistical models for example the Average of Rates (AOR), Least Median of Squares (LMS), Linear Regression Rate (LRR), End Point Rate (EPR) model, and Jackknife model (JK) were used to evaluate shoreline prediction [12, 14–19].

Few encouraging investigations have been conducted along Egypt's North Sinai shore [20] used an aerial picture taken in 1955 and a topographic map analyzed in 1992 to describe the shoreline alteration along Sinai's northern coast. Despite this, the magnitude of shoreline variations was not quantified in their analysis due to the inability of the analyzed maps' surveying methodologies to calculate it. Moreover [21] used a hydrographic survey to investigate the impact of the El Arish power station (located west of the El Arish valley coast) on the surrounding area on Sinai's Mediterranean coast. The authors discovered a 5.5 m/year coastline retreat east of the El Arish power plant breakwater [22] used topographic maps from 1973 with satellite pictures from 1984 and 1996 to tutor the coastal changes over the western half of the North Sinai coast (i.e., from El Tinah Bay to El Bardawil Lake). They also calculated how the area of El Bardawil Lake changed throughout time. They discovered that the extent of El Bardawil Lake changed dramatically from 1973 to 1984, losing an average of 128 km^2, then slowing to a loss of 5 km^2 from 1984 to 1996. El Banna et al. (2009) used the same method to track changes in the shoreline along the North Sinai coast for 15 years (from El Bardawil Lake to Rafah) by studying TM and ETM true color Landsat pictures from 1986 to 2001. The accretion and erosion rates were calculated to be +0.076 km^2 $year^{-1}$ and 0.123 km^2 $year^{-1}$, respectively.

To sum up, the extensive literature survey undertaken in the North Sinai coast area demonstrated that the current published data related to the area needs to be improved and renovated. Furthermore, it cannot determine coastline change rates with high-precision approaches. The current study uses GIS and DSAS geospatial approaches to examine shoreline changes along the North Sinai coastline from 1989 to 2016. Furthermore, the current research aims to: (1) apply three different semi-automated shoreline extraction methods, including Histogram threshold of band ratio, Histogram threshold of band 5, and Tasseled Cap Transformation (TCT); (2) plot and measure shoreline accretion/erosion rates using several statistical methods functionalized in DSAS, including NSM, LRR, EPR, and LMS; and (3) develop a decision-support algorithm that can vigorously support in elaborating shoreline accretion/erosion rates; (4) using the EPR model, outline a futuristic decision plotting based on the North Sinai shoreline forecast in the years 2025, 2035, and 2050.

2. Study area

Sinai's coastal area is considered an essential part of Egypt's Mediterranean Coast [22]. It is a geographical connecting point between Asia and Africa, with the Gulf of Suez and the Suez Canal on the west, the Gulf of Aqaba and the Egyptian-Israeli border on the east, and the Mediterranean Sea on the north (**Figure 1**). The latitudes and longitudes are (28°–31°N) and (32° 30′–34° 30′ E) respectively. The northern Sinai coast stretches for about 220 km along the Mediterranean Sea, extending from Port Said in the west to Rafah in the east, from the Egyptian border [23]. The current

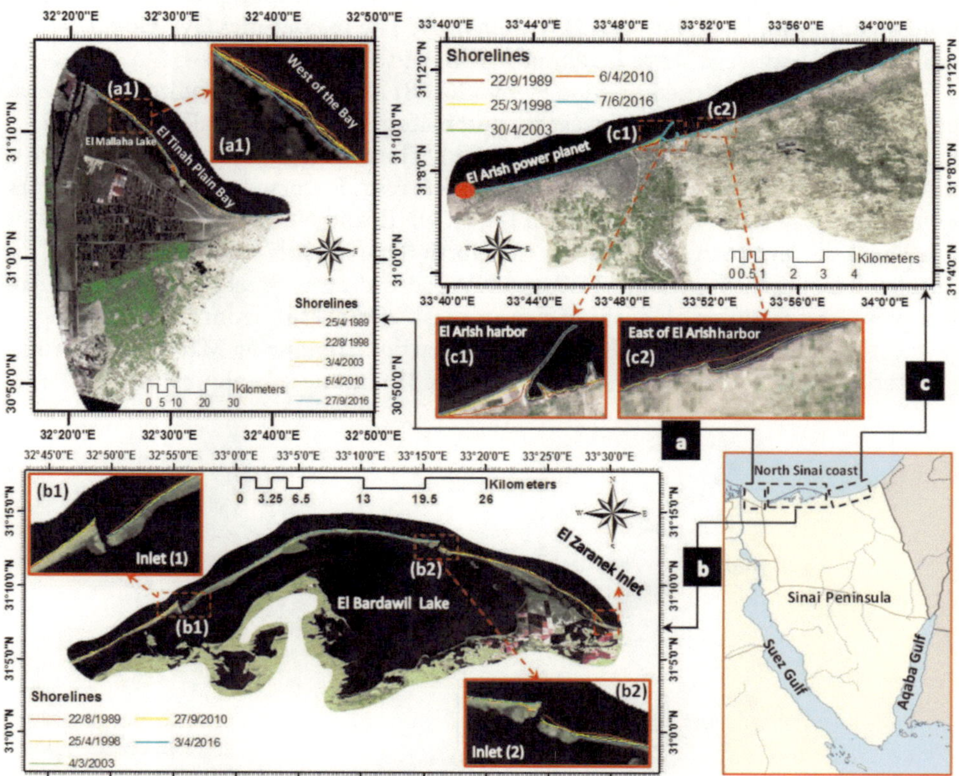

Figure 1.
False-color composite images of the study area and shoreline digitization in different periods from (1989–2016) for the three zones (a, b, c) respectively, after [4].

study area is split into three subzones based on the vulnerability of coastal areas as well as the availability of data from the field and remotely sensed data. Zone I contain El Tinah Plain Bay, which stretches 38.5 km from Port Said in the west to El Bardawil Lake in the east, (**Figure 1a**). Zone I is characterized by some features such as Lagoons, vegetation cover, and fish ponds. Zone II includes El Bardawil Lake (**Figure 1b**). This lake covers approximately 60% of Sinai's northern coast. It has a total area of over 700 km^2 and is approximately 72.5 km long, 22 km wide-ranging, and 2 m deep. The Mediterranean Sea is isolated from the Lake by shallow sand barriers that range in width from 300 to 1000 m, and are overtopped by storm waves in the winter. It has three inlets joining it to the Mediterranean Sea, two of which are manmade (no. 1 and 2) and one of which is natural (El Zaranek inlet), [24]. Zone III includes the El Arish Valley shore, which almost forms a 37 km west-to-east intersection between the El Arish power plant and Rafah, (**Figure 1c**).

2.1 Wave climate at study area

The intensity and direction of wave action along Egypt's Mediterranean coast are inextricably linked to significant pressure systems over the Mediterranean and North Atlantic [25]. Wave heights reach 1.16 m and average 0.4 m during the spring and summer, with the prevailing wave direction being NW. Prevailing wave direction is come mainly from N, NNW, and NW in Winter. The maximum wave height is 4.25 m, with an average wave height of 0.51 m and a period of 6.5 sec.

Wave data was analyzed previously by [26–28] along Egypt's Mediterranean coast show that waves from the northwest predominate (81%), with small components from the northeast (14%) and southwest (5%). The prevailing wave direction is the main key of the eastward-flowing alongshore current. Reversed alongshore currents are generated by waves incoming from the N, NNE, and NE (**Figure 2**). The main alongshore current path (62–65%) on the North Sinai coast is from west to east, stimulated by waves from the NNW and NW, according to preceding measured data. However, west trending alongshore currents (24–29%) result from the remaining wave components from the N, NNE, and NE, particularly during March and April due to easterly winds. Furthermore, with a range of 31 cm, the tide along Sinai's

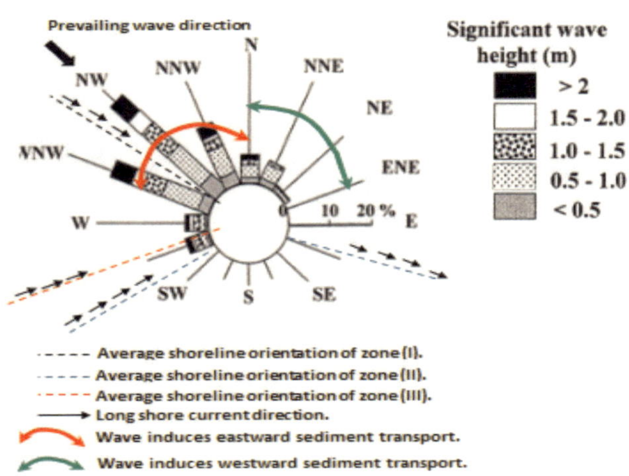

Figure 2.
Wave rose for study area, and wave induced currents' directions (modified from El Banna et al., 2009), after [29].

Mediterranean coast is micro-tidal and semi-diurnal. The average high water level is 20 cm, and the average low water level is −11 cm [30].

3. Materials and methods

3.1 Data source

This study used multi-temporal satellite data from Landsat TM, ETM, and OLI/TIRS that cover our coast from 1989 to 2016. Even so, thanks to the shortage of cloud-free imagery during the selected period, satellite images could not be obtained at regular intervals. The polynomial geo-rectification method is used to ortho-rectify the selected satellite images, as it is afterward used to track changes in the shoreline along the Sinai Peninsula's northern coast Satellite images' data are described in detail in **Table 1**. Data acquired for North Sinai coastline surveying from El Tinah bay to El Arish valley was supplied by the Egyptian Institute of Oceanography and Fisheries in 2010.

3.2 Image processing

Image processing carried out in this study were strip filling, georeferencing, and radiometric correction. Firstly, gap filling was applied to image 2010 for all its bands using modeling done by [31] in Arc GIS 10.2.2 using the python algorithm, see **Figure 3**.

Ground Control Points (GPCs) are used to implement the geometric correction process (i.e. more than 40 GCPs are identified on the images), [10, 32]. The geometric correction is accomplished utilizing ENVI 5.3 software to reduce distortions caused by scale variation, angle, and lens distortion.

Satellite data	Path/Row	Year of acquisition	Resolution (pixel size)	Zone
Landsat 4 - TM	176/38	1989	30 m	I
Landsat 5 - TM	175/38	1989	30 m	II and III
Landsat 5 - TM	176/38	1998	30 m	I
Landsat 5 - TM	175/38	1998	30 m	II and III
Landsat 7 - ETM	176/38	2003	30 m	I
Landsat 5 - TM	175/38	2003	30 m	II and III
Landsat 7 - ETM	176/38	2010	30 m	I
Landsat 7 - ETM	175/38	2010	30 m	II and III
Landsat 8 - OLI/TIRS	176/38	2016	15 m	I
Landsat 8 - OLI/TIRS	175/38	2016	15 m	II and III
Related abbreviations:				
TM	: Thematic Mapper;			
ETM	: Enhanced Thematic Mapper;			
OLI	: Operational Land Imager;			
TIRS	: Thermal Infrared Sensor.			

Table 1.
Details of satellite dataset for North Sinai (acquired via https://earthexplorer.usgs.gov/).

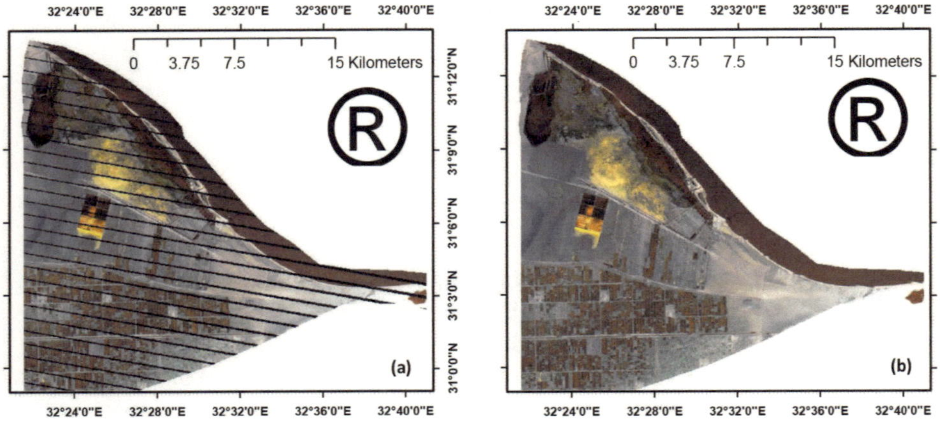

Figure 3.
Landsat image for zone I in 2010; (a) before gap filling; (b) after gap filling.

The image is projected to actual coordinate Universal Transverse Mercator (UTM), WGS-84 datum. After georeferencing, (RMSE) was found to be less than 0.5 pixels, indicating that the images were geometrically well-matched. After that, a radiometric correction is applied using the ENVI software's radiometric, which combines the sun and view angle effects, as well as sensor calibration and atmospheric correction. Eventually, all georeferenced images are processed in ArcGIS to get the coastline digitized.

3.3 Shoreline delineation and uncertainties

Shorelines are the high water line as surveyed by GPS units in kinematic mode [33]. Meanwhile, automatic coastline demarcation from low resolution satellite images is a complicated job due to the unclear boundary between water and land in saturated zone [34]. Three semi-automatic delineation approaches are first tried for Landsat images ETM 2010 in this study to identify the best digitization methodology that gives the least error with the related field data in 2010 (**Figure 4**).

Since water absorbs the majority of radiation in the near-infrared and mid-infrared regions of the spectrum, its reflectance in these wavelengths is nearly zero; nevertheless, the reflectance is higher in these areas for land cover than water bodies. As a result, the coastline can be derived from a single band image. As a consequence, getting the binary image is becoming simple by estimating the histogram threshold for one of the infrared bands (i.e. Band 5) of the TM or ETM imagery [35]. Another method is to use the histogram threshold of band ratio technique, which produces a binary image by combining the two conditions of Band (2)/Band (4) $\geq$ 1.0 and Band (2)/Band (5) $\geq$ 1.0 for producing a binary image [36].

Moreover, the Tasseled Cap Transformation technique (TCT) is also used to extract shorelines. The coefficients for TCT of Landsat data are determined from [37]. TCT reconstitutes the spectral information of the six ETM bands into three primary perspective elements using coefficients deduced from sampling known land cover spectral features. The moistness component is used to distinguish land from water among the three main view elements (brightness, greenness, and wetness). In this

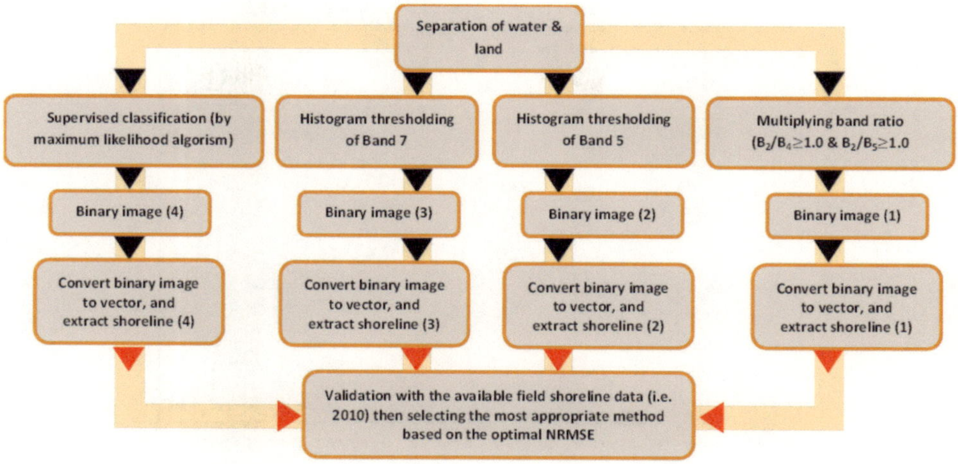

Figure 4.
Methodology framework to extract shoreline.

wetness index band, the land-water configuration is clearly visible, and a binary image could be easily acquired. Finally, in 2010, each technique's raster binary image is transformed into a vector image, which can then be used to extract the coastline border.

To identify the best technique, a comparison is made in Arc GIS 10.2.2 for the three regions between the derived shorelines (e.g., by TCT) and the observed shorelines in 2010 (**Figure 5**). This comparison is assessed using DSAS tools, which developed by the United States Geological Survey (USGS). The original purpose of this extension is to compute change rate in coastline' positions. Using DSAS is summarized in the following steps: After extraction of shoreline, the baseline is created; creating transects; calculating the distances between coastline and baseline for transects; finally, the shoreline change rate is calculated [38]. Accordingly, the deviation between the derived and observed shoreline is computed using 1800 perpendicular to the baseline transects. These transects are accurately cast at intervals of 20 m (**Figure 5a–c**).

Figure 6 depicts a validation process between data from a field investigation and extracted shoreline from satellite image obtained in 2010. It is based on the coupling of DSAS software and Arc GIS 10.2.2. The residuals between the measured and computerized shorelines in 2010 at each transect line from 1 to 1800 were estimated using both the histogram threshold of band 5, histogram threshold of band ratio, and TCT, as shown in **Figure 6a–c**. It is noticed that data are reasonably correlated (**Figure 6a1**, **b1**, and **c1**).

The normalized root means square error (NRMSE) is considered to find the best method that precisely extract the coastline. TCT technique is proved to be better in shoreline delimitation using low resolution satellite imagery (medium resolution). It achieved the least NRMSE for all zones, **Figure 6a1**, **b1**, and **c1**. As a result, the TCT technique was used to demarcate shorelines in 1989, 1998, 2003, and 2016 (see sectors a1, b1, b2, c1, and c2 in **Figure 1a–c**).

3.4 Shoreline 'change rate

Changes in the shoreline locations are calculated using different four analysis methods (i.e., EPR, LRR, LMS, and NSM). The End Point Rate (EPR) is easily

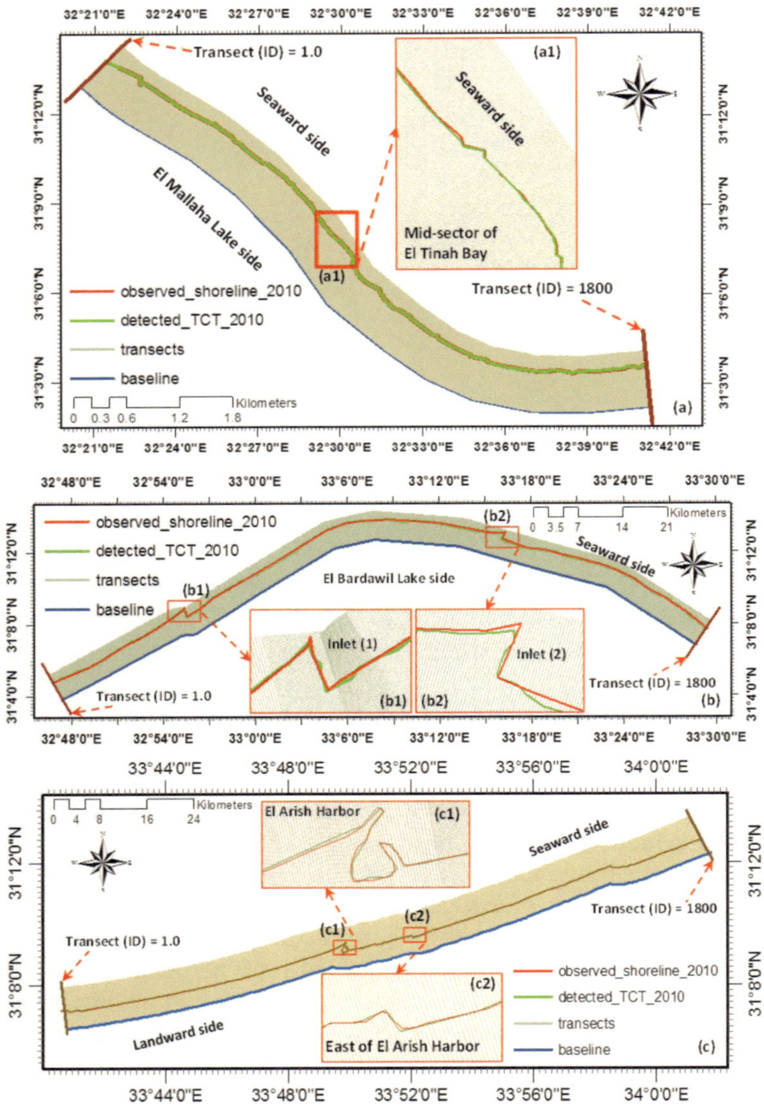

Figure 5.
Comparison of digitized shorelines based on field data from 2010 and the corresponding Landsat imagery (e.g., using TCT) for (a) Zone (I); (b) Zone (II); (c) Zone (III).

determined by dividing the length (in m.) between two coastlines by number of the years (Eq. (1) and **Figure 7**). This method is widely used by different coastal researchers and is widely used in shoreline movement rate calculations [39–42].

$$EPR = \frac{L_1 - L_2}{t_1 - t_2} \tag{1}$$

where:

L_1 and L_2 are the distances between the baseline (benchmark) and the shoreline, while t_1 and t_2 are the dates of the two shoreline locations.

Linear Rate Regression (LRR) is the second method for calculating change rates. For a specific transect, this method entails fitting a least-squares regression line to

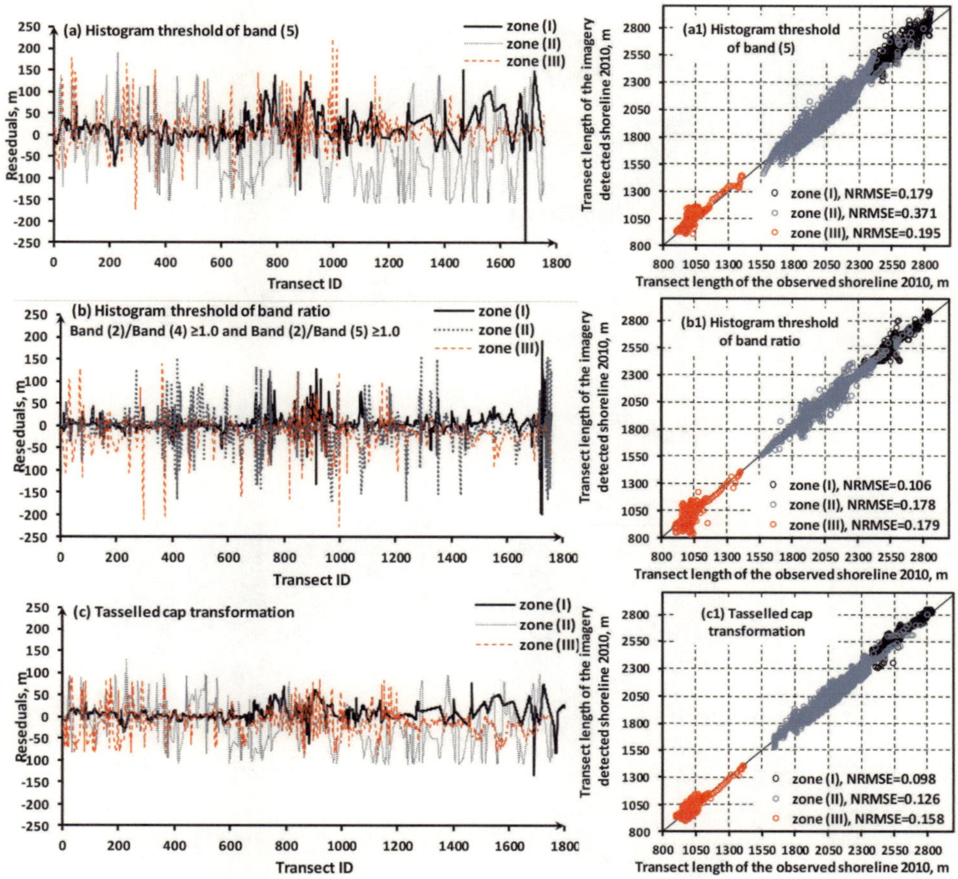

Figure 6.
Validation process between the shoreline monitored in the field and the shoreline detected by imagery 2010 for the different zones based on NRMSE of ; (a,a1) histogram threshold of the band (5); (b,b1) histogram threshold of band ratio; (c,c1) TCT.

multiple shoreline location points, (**Figure 8**). R-squared (Eq. (2)), $R^2 > 0.87$ has been held as the threshold of certainty in our research, considering a confidence interval (LCI) of 95%. R^2 at each transect line are calculated as follows:

$$R^2 = 1 - \sqrt{\frac{\sum_{i=1}^{N}\left(L - L_p\right)^2}{\sum_{i=1}^{N}\left(L - L^{-}\right)^2}} \tag{2}$$

where:

L: observed distance between the reference line(baseline) for a coastline' data point;

L_p: forecast value based on the best-fit linear regression equation;

L^{-}: Average of the observed shoreline data points; and.

N: number of dates.

The sample data are used to calculate an average offset in the linear regression method, and the formula for the line is deduced by reducing this value so that the source points are as near to the regression line as possible. In the least median of

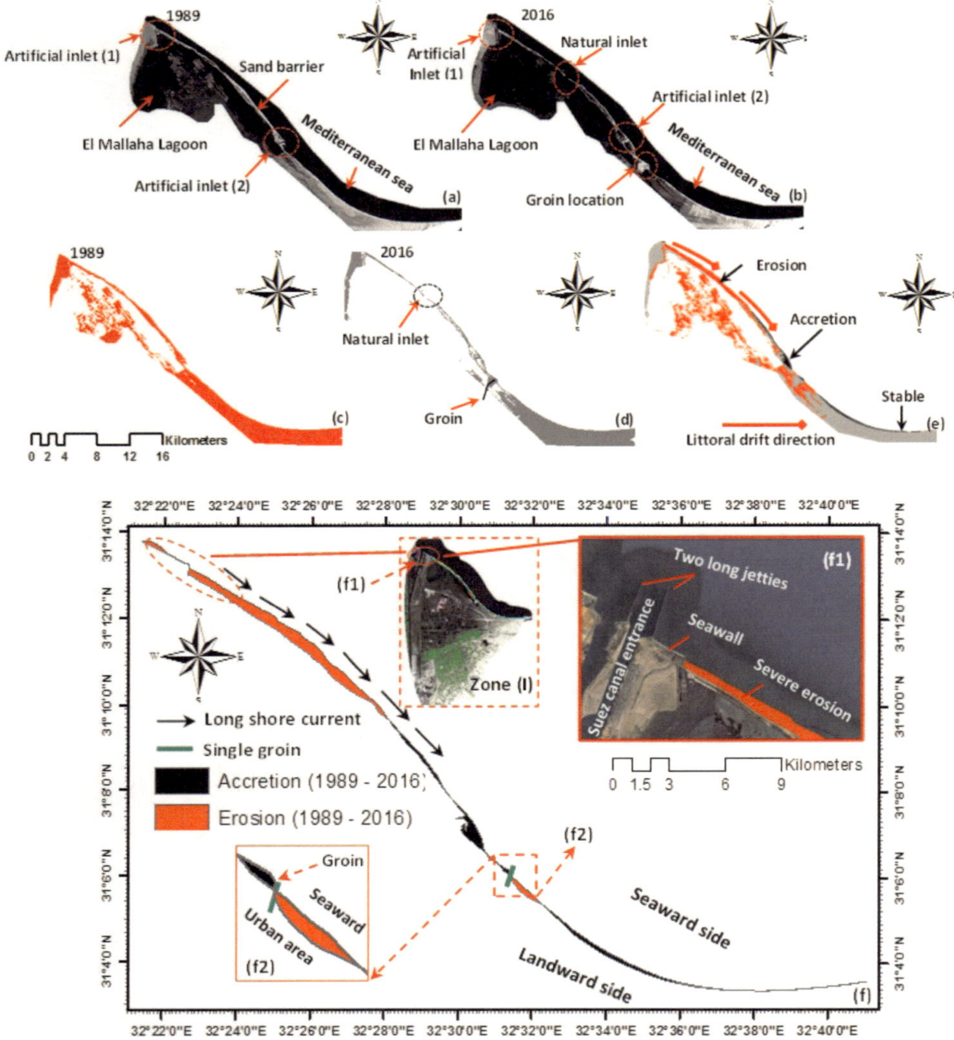

Figure 7.
Detecting changes of zone I; (a, b) Satellite images (TM) and OLI/TIRS of band (5) for the year1989 and 2016; (c, d) TCT's equivalent binary images from 1989 and 2016; (e) The Change detection image for the period 1989 to 2016; (f) Vector map showing the erosion/accretion pattern showed in vector map for the period 1989 to 2016.

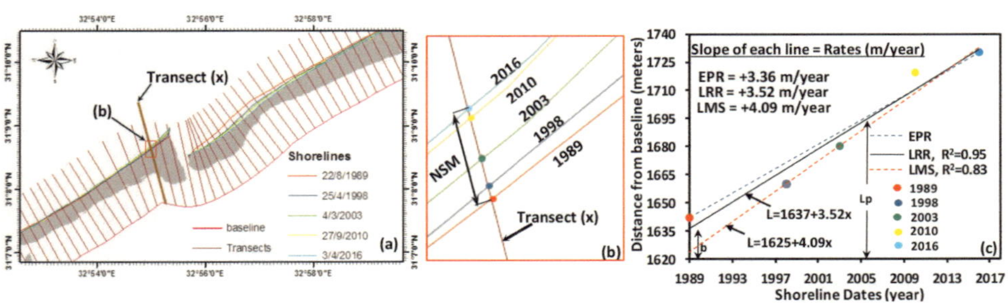

Figure 8.
Explanatory example of NSM, EPR, LRR, and LMS computation; (a) Map of multi-temporal shoreline locations west and east El Bardawil inlet (1); (b) transect line' details (x) and coastline intersection; (c) Time series of shoreline distances from the baseline along the transect line (x).

Category	shoreline change' rate (m/year)	classification of shoreline
1	> -2	Very high erosion
2	> -1 and < -2	High erosion
3	> 0 and < -1	Moderate erosion
4	0	Stable
5	> 0 and < +1	Moderate accretion
6	> +1 and < +2	High accretion
7	> +2	Very high accretion

Table 2.
The classification of Shoreline according to EPR, LRR, and LMS.

squares method (LMS), instead of using the average, the median value of the squared residuals is used to identify the optimal equation for the line (**Figure 8**).

The net spacing (in meters) between the past and present shoreline locations for each transect is recognized as the Net Shoreline Movement (NSM) (i.e., 1989 and 2016). It represents a distance rather than a rate (**Figure 8b**).

EPR, LRR, LMS, and NSM have negative values, implying landward decline of the shoreline, whereas positive values indicate landward advancement. The erosion/ accretion rates measured along the North Sinai coast are divided into seven categories (**Table 2**) [43].

4. Results and discussion

4.1 Historical shoreline change detection (1989–2016)

A long-term process of two-dimensional shoreline change detection has been extensively investigated along the coastal line of different three zones over a 27-year period (1989–2016). This procedure is conducted through different steps. Firstly, binary images from 1989 and 2016 are derived for each zone using TCT techniques to separate land and water. This step masks the land cover with all of its categories. Second, the binary images are converted from raster to vector (feature class) using ArcGIS10.2.2 software, with two main polygon attributes: water and land. Finally, the two polygon layers are superimposed to assess shoreline erosion/accretion trend from 1989 to 2016, (**Figure 7f**).

As a result, **Figure 7a** and **b** show satellite TM and OLI/TIRS images of the band (5) for zone I in 1989 and 2016, respectively, while **Figure 7c** and **d** show their classified binary images The post-classification change detection image (**Figure 7e**) on the other hand, shows severe erosion in El Tinah Bay's western part. This erosion is the result of the combined effects of the coast's stormy climate and the restriction of sediment movement from the Nile Delta as a result of the construction of both the jetties at Suez Canal entrance and seawalls at eastern canal. Besides that, a portion of the incident wave's energy is shifted into the adjacent beach due to the construction of this seawall. Consequently, the shifted energy, soil disconnection has occurred in the western part of El Tinah Bay. Based on the hydrodynamic processes on the North Sinai coast, alongshore currents induced sediments to move from west to east, resulting a highly sensitive eroded area (**Figure 7f, f1**).

In 2013, a natural opening nearly in the middle of El Mallaha Lagoon was formed as a result of this erosion. Furthermore, as a result of hindering the sediment path by inlet (2) jetties and groins at east of the inlet, part of the transported sediments has settled nearly in the middle of El Tinah Bay shoreline. In 2015, the artificial inlet (2) was completely blocked due to sediment restrictions.

The eastern part of the Bay, on the other hand, appears to be relatively stable. This part's shoreline is almost straight, with no major merged parts to erode or major embayments to receive sediments, hence, the shore zone is nearly stable; Only a few small pockets of accumulation have been noticed. To quantify the dynamical changes in the zone I coastline from 1989 to 2016, an asymmetrical difference vector map is created from binary images in ArcGIS10.2.2 and is then classified into two categories: erosion and accretion pattern, (**Figure 7f**). The change detection clearly shows a cumulative accretion of +3.442 km^2 and a rate of +0.127 km^2/year, while the cumulative erosion is −5.409 km^2 and a rate of −0.2 km^2/year over a period of 27 years (**Table 3**).

The defined trend of the two-dimensional shoreline change rate (km^2/year) along the coastline, extracted using TCT technique is noticed to be rather coherent with other earlier studies when compared to the other two remote sensing techniques. This is evident when the current results have been compared with the previous results in researches of [22, 23] as shown in **Table 3** and **Figure 9**. Moreover, owing to variation

Study zone	Research	Date	Data source	The used technique	Area (km^2)			Rate (km^2/year)		
					Loss	Gain	Net	Loss	Gain	Net
I	(Azab & Noor 2003)	1973-1996	Topographic maps	Manually by FCC	-3.22	2.42	-0.81	-0.14	0.105	-0.035
	Present study	1989-2016	Satellite imagery	TCT	-5.41	3.44	-1.97	-0.2	0.127	-0.073
II	(Azab & Noor 2003)	1973-1996	Topographic maps	Manually by FCC	-4.12	2.19	-1.93	-0.179	0.095	-0.084
	(El Banna et al. 2009)	1986-2001	Satellite imagery	Histogram threshold	-3.03	1.52	-1.52	-0.202	0.101	-0.101
	Present study	1989-2016	Satellite imagery	TCT	-6.95	4.37	-2.57	-0.257	0.162	-0.095
III	(El Banna et al. 2009)	1986-2001	Satellite imagery	Histogram threshold	-0.63	0.77	0.138	-0.042	0.051	0.0092
	Present study	1989-2016	Satellite imagery	TCT	-1.61	1.947	0.339	-0.059	0.072	0.0126
Overall	Present study	1989-2016	Satellite imagery	TCT	-24.9	16.6	-8.31	-0.925	0.61	-0.308

Table 3.
Calculated area of erosion(-ve), accretion(+ve), and net balance surfaces along the North Sinai coast.

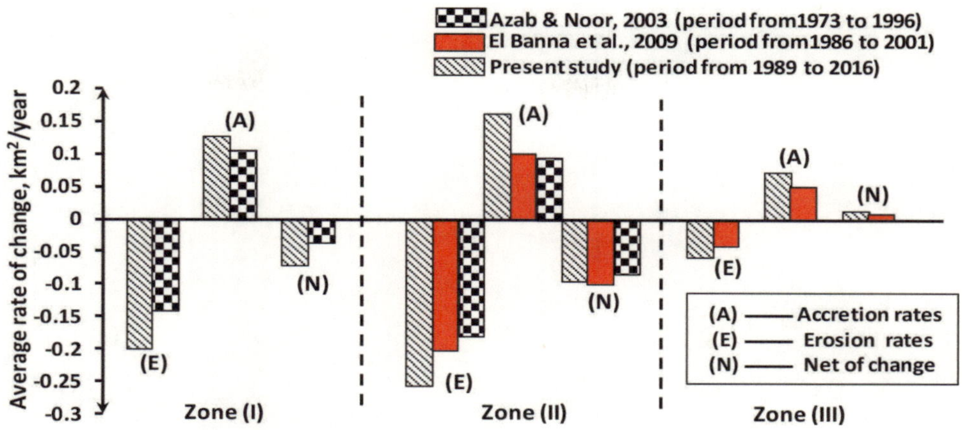

Figure 9.
Mean change rates (km²/year) along the North Sinai shoreline.

in both the picked elapsed times for each research and the accuracy of each methodology, there is a quiet discrepancy between the results.

4.2 Analysis of shoreline kinematics by DSAS

The digitized shorelines have been used in the ArcGIS extension Digital Shoreline Analysis System (DSAS) to calculate the rate of shoreline change in vector format over a specific period of time [44]. DSAS is a statistical software applied in coastal research to compute rate of change from historical GIS-based shoreline positions [45].

In this study, DSAS is utilized to calculate shoreline change rate from different historical shoreline locations along the North Sinai coast in 1989, 1998, 2003, 2010, and 2016. The method for determining shoreline change rates begins with the creation of a personal geodatabase for the extracted shoreline positions in ArcCatalog 10.2.2. Each shoreline has attributes that include date, length, ID, shape, and uncertainty. Each image's acquisition date is entered in the date column, whereas the length, ID, and shape are easily obtained. Uncertainties are also measured (**Table 2**) and recorded in the uncertainty column as integers. The five shoreline positions are then appended to one shapefile. Thereafter, speculative baseline is formed from the shoreline. Three different methods are available in DSAS to delineate baseline: (1) constructing a baseline along the shoreline at a particular distance; (2) utilizing a previously established baseline; (3) buffering method. The last method is the most consistent and accurate technique for baseline demarcation because it uses the same sinuosity shape as the nearby shoreline, so it was selected for the current study [12].

The baseline is then created at a buffering distance of 1000 meters offshore from the nearest shoreline.

These attributes provide DSAS with information about the sequence of transects as well as the baseline's position in relation to the shoreline (onshore or offshore). Transects have been set orthogonally from the benchmark (baseline) along the coastline of various years in 100 m intervals for the three different zones. Finally, the shoreline change rates are statistically computed using the various techniques (i.e., EPR, LRR, LMS, and NSM). As shown in (**Figure 10a–c**), a qualitative analysis is performed to determine the related erosion/accretion transects using the NSM model.

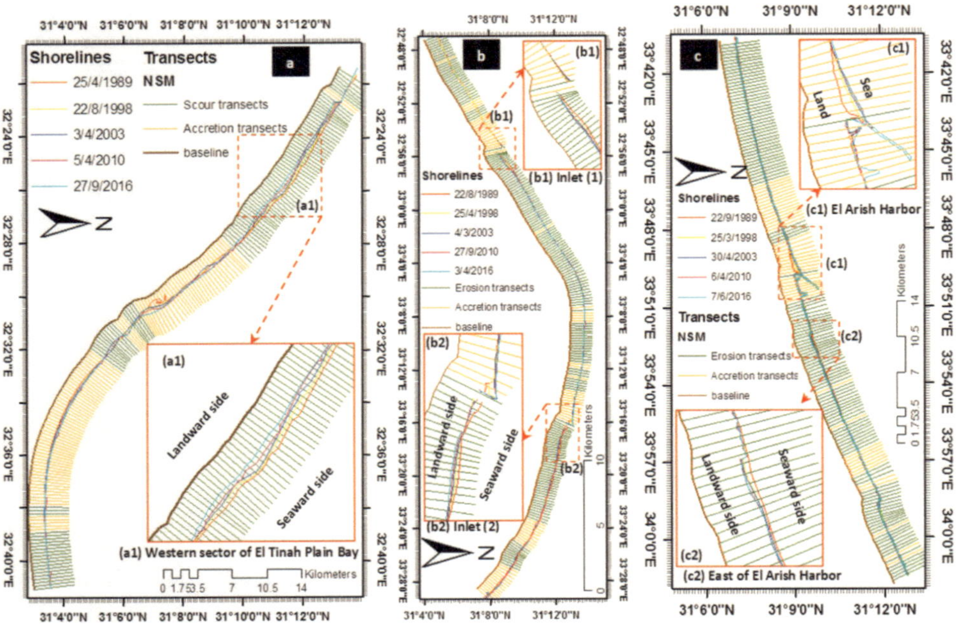

Figure 10.
Qualitative analysis of erosion/accretion transects using NSM, conducted in DSAS, in the years of 1989, 1998, 2003, 2010, and 2016 for (a) zone (I); (b) zone (II); (c) zone (III).

The field that connects the table of NSM statistical results to the transect feature class is the field that they have in common. Where the values in the transect-ID field of the NSM results table are equal to the object identifier field (Object ID) in the transect feature class. After completing the joining process, the symbology of the transect feature class can be adjusted to classify transects into two categories: erosion (green transects) and accretion (orange transects), (**Figure 10**). Most beaches in zones I, II, and III are susceptible to accretion and retreat (1989–2016), according to the delineation of erosion and accretion transects.

Furthermore, the results reveal that, 49.61% (191 transects), 73.52% (533 transects), and 72.24% (255 transects) of the coastline corresponding to 19.1, 53.3, and 25.65 km are experiencing erosion for zone I, II, and III respectively. On ht. eother hand, 50.39%, 26.48%, and 27.76% of the coastline with lengths of 19.4, 19.2, and 9.85 km are suffering accretion (**Table 3**).

The rates of shoreline change are computed annually for each zone during the period from 1989 to 2016 using the statistical outcomes of EPR, LRR, and LMS (**Figure 11**). The findings of this study are summarized in **Table 3**, which shows the average and maximum rates of coastline advancement and decline, as well as the percentage of degradation and deposition transects, both locally and globally. In **Figure 11** and **Table 3**, the positive and negative values of the EPR, LRR, and LMS reveal accumulation and recession areas, respectively. Apart from the last quarter of the eastern part of the El Tinah plain Bay coastline, which appears to be markedly stable over a quarter-centennial period (27 years), almost the entire coastal area of the North Sinai is sensitive to either accretion or retreat, as shown in **Figure 11a–c**. Meantime, when comparing the outputs of EPR and LRR to the outputs of LMS, the overall measurable change rates of the North Sinai coast show a significant affinity. Whereas, the achieved R-squared values show a strong correlation

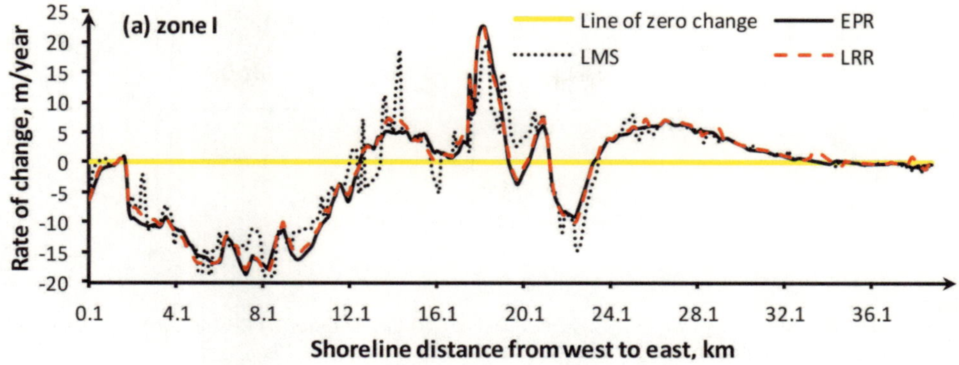

Figure 11 Shoreline change rates by EPR, LRR, and LMS (m/year) for Zones, (a) I; (b) II; and (c) III during the period from 1989 to 2016.

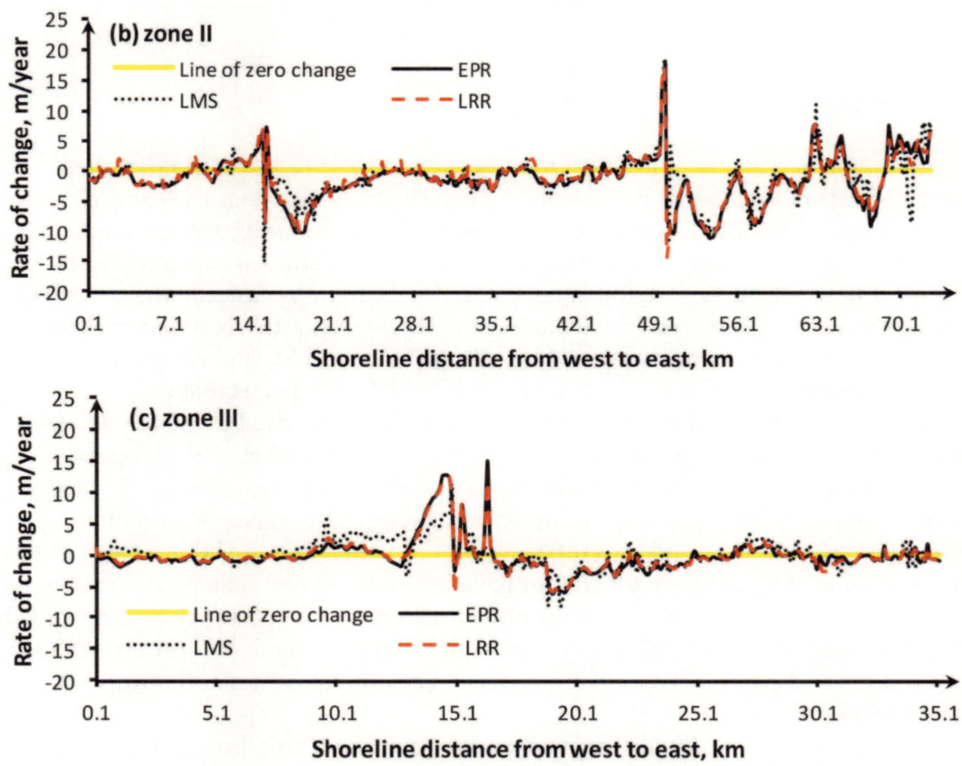

Figure 11.
Shoreline change rates by EPR, LRR, and LMS (m/year) for Zones, (a) I; (b) II; and (c) III during the period from 1989 to 2016. Shoreline change rates by EPR, LRR, and LMS (m/year) for Zones, (a) I; (b) II; and (c) III during the period from 1989 to 2016.

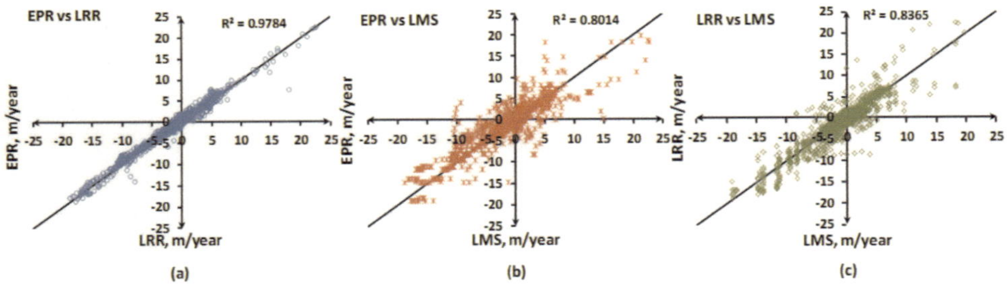

Figure 12.
Comparison of shoreline change rates (m/year) calculated by, (a) EPR vs LRR; (b) EPR vs LMS; (c) LRR vs LMS for the overall North Sinai coast.

between EPR and LRR, with a value of 0.978. The situation is different regarding EPR vs. LMS and LRR vs. LMS, as the correlation values are 0.8 and 0.836 respectively (**Figure 12a–c**).

5. Conclusions

Geospatial techniques and DSAS models were utilized to assess the shoreline morphodynamic changes along the North Sinai shoreline between 1989 and 2016 via multi-temporal satellite images. The semi-automatic shoreline extraction method (Tasseled Cap Transformation technique, TCT) was accustomed to digitalize the shoreline positions in 1989, 1998, 2003, and 2016. Extreme variance in the spatial scale characterizes the study area where the highest obtained coastal erosion/accretion kinematics for El Arish valley coast, El Bardawil Lake, and El Tinah Bay are $-1.61/+1.95$ km^2, $-6.95/+4.37$ km^2, and $-$ 5.41/+3.44 km^2, respectively.

Moreover, the construction of eastern jetty of the Suez Canal extremely lowered sediments inputs to El Tinah Bay, which highlighted the erosion of the western segment by wave hydrodynamics and the eastwards alongshore currents. Contrary, the eastern part of El Tinah Bay has demonstrated a nearly constant shoreline throughout the study period. Instantaneously, protection jetties of El Bardawil inlet 1, El Bardawil inlet 2, and El Arish Harbor have intermittent long-shore sand movement resulting in a continuous erosion at the downdrift side and an accretion at their updrift side. The institution of El Arish power plant has substantially decreased the sedimentary routine and created destructive impacts on coastal dynamics in the west of El Arish harbor.

In the meantime, the forthcoming speculation of the North Sinai coastline variations is predicted using the End Point Rate (EPR) model for the near future of years 2025, 2035, and 2050 after model validation based on 2010 data. A well-matching between the historical and futuristic trends of shoreline is obtained which means the calculation is almost succeeding the same accretion and erosion patterns. Study results in this chapter prove that medium-resolution satellite images, geospatial features of the GIS, and digital shoreline analysis system (DSAS) successfully assessed the coastal morphodynamic changes and shoreline detection of the North Sinai coast and could be used for other coastal areas based on the data quality and availability. Additionally, the results of this study deliver a high-reliable tool to the decision-makers and the coastal managers to support their decision when developing sustainable coastal management plans for North Sinai coast.

Acknowledgements

The author would like to thank the editor and the reviewers for their constructive comments for enhancing the chapter quality.

Author details

Ali Masria[1,2*], Karim Nassar[2] and Mohamed Galal Eltarabily[3,4]

1 Civil Engineering Department, College of Engineering, Jouf University, Saudi Arabia

2 Irrigation and Hydraulics Engineering Department, College of Engineering, Mansoura University, Mansoura, Egypt

3 Civil Engineering Department, Faculty of Engineering, Port Said University, Port Said, Egypt

4 UC Kearney Agricultural Research and Extension Center, University of California, Parlier, CA, USA

*Address all correspondence to: aatef@ju.edu.sa; ali_maaasria@mans.edu.eg

References

[1] Addo KA, Walkden M, Mills JPT. Detection, measurement, and prediction of shoreline recession in Accra, Ghana. ISPRS Journal of Photogrammetry and Remote Sensing. 2008;**63**(5):543-558

[2] Zhang Y, Lu D, Yang B, Sun C, Sun M. Coastal wetland vegetation classification with a Landsat Thematic Mapper image. International Journal of Remote Sensing. 2011;**32**(2):545-561

[3] Genz AS, Fletcher CH, Dunn RA, Frazer LN, Rooney JJ. The predictive accuracy of shoreline change rate methods and alongshore beach variation on Maui, Hawaii. Journal of Coastal Research. 2007;**23**(1):87-105

[4] Van TT, Binh TT. Shoreline change detection to serve sustainable management of coastal zone in Cuu Long Estuary. In: International Symposium on Geoinformatics for Spatial Infrastructure Development in Earth and Allied Sciences. Vol. 1. 2008

[5] Appeaning Addo K, Jayson-Quashigah PN, Kufogbe KS. Quantitative analysis of shoreline change using medium resolution satellite imagery in Keta, Ghana. Marine Sciences. 2012;**1**(1): 1-9. DOI: 10.5923/j.ms.20110101.01

[6] Kroon A, Kabuth AK, Westh S. Morphologic evolution of a storm surge barrier system. Journal of Coastal Research. 2013;**65**:529-534. DOI: 10.2112/SI65-090

[7] Tran Thi V, Xuan ATT, Nguyen HP, Dahdouh-Guebas F, Koedam N. Application of remote sensing and GIS for detection of long-term mangrove shoreline changes in Mui Ca Mau, Vietnam. Biogeosciences. 2014;**11**(14): 3781-3795. DOI: 10.5194/bg-11-3781-2014

[8] Murali RM, Dhiman R, Choudhary R, Seelam JK, Ilangovan D, Vethamony P. Decadal shoreline assessment using remote sensing along the central Odisha coast, India. Environment and Earth Science. 2015;**74**(10):7201-7213. DOI: 10.1007/s12665-015-4698-7

[9] El Sharnouby BA, El Alfy KS. Coastal changes along Gamasa Beach, Egypt. Journal of Coastal Zone Management. 2015;**18**(1). DOI: 10.4172/2473-3350.1000393

[10] Masria A, Nadaoka K, Negm A, Iskander M. Detection of shoreline and land cover changes around Rosetta Promontory, Egypt, based on remote sensing analysis. Landscape. 2015;**4**(1): 216-230. DOI: 10.3390/land4010216

[11] Bheeroo RA, Chandrasekar N, Kaliraj S, Magesh NS. Shoreline change rate and erosion risk assessment along the Trou Aux Biches–Mont Choisy beach on the northwest coast of Mauritius using GIS-DSAS technique. Environment and Earth Science. 2016;**75**(5):1-12. DOI: 10.1007/s12665-016-5311-4

[12] Nandi S, Ghosh M, Kundu A, Dutta D, Baksi M. Shoreline shifting and its prediction using remote sensing and GIS techniques: A case study of Sagar Island, West Bengal (India). Journal of Coastal Conservation. 2016;**20**(1):61-80

[13] Kermani S, Boutiba M, Guendouz M, Guettouche MS, Khelfani D. Detection and analysis of shoreline changes using geospatial tools and automatic computation: Case of jijelian sandy coast (East Algeria). Ocean and Coastal Management. 2016;**132**:46-58. DOI: 10.1016/j.ocecoaman.2016.08.010

[14] Burgess K, Jay H, Hosking A. Futurecoast: Predicting the future

coastal evolution of England and Wales. Journal of Coastal Conservation. 2004;**10** (1–2):65-71. DOI: 10.1652/1400-0350 (2004)010[0065:FPTFCE]2.0.CO;2

[15] Kuleli T. Quantitative analysis of shoreline changes at the Mediterranean coast in Turkey. Environmental Monitoring and Assessment. 2010;**167** (1–4):387-397. DOI: 10.1007/s10661-009-1057-8

[16] Chand P, Acharya P. Shoreline change and sea-level rise along coast of Bhitarkanika wildlife sanctuary, Orissa : An analytical approach of remote sensing and statistical techniques. International Journal of Geomatics Geoscience. 2010;**1**(3):436-455

[17] Mukhopadhyay A, Mukherjee S, Mukherjee S, Ghosh S, Hazra S, Mitra D. Automatic shoreline detection and future prediction: A case study on Puri coast, Bay of Bengal, India. European Journal of Remote Sensing. 2012;**45**(1): 201-213. DOI: 10.5721/EuJRS20124519

[18] Mondal I, Bandyopadhyay J, Dhara S. Detecting shoreline changing trends using principle component analysis in Sagar Island, West Bengal, India. Spatial Information Research. 2017;**25**(1):67-73. DOI: 10.1007/s41324-016-0076-0

[19] Bruno DE, Barca E, Goncalves RM, de Araujo Queiroz HA, Berardi L, Passarella G. Linear and evolutionary polynomial regression models to forecast coastal dynamics: Comparison and reliability assessment. Geomorphology. 2018;**300**:128-140. DOI: 10.1016/j.geomorph.2017.10.012

[20] Frihy OE, Lotfy MF. Shoreline changes and beach-sand sorting along the northern Sinai coast of Egypt. Geo-Marine Letters. 1997;**17**(2):140-146. DOI: 10.1007/s003670050019

[21] Frihy OE, Badr AA, Selim MA, El Sayed WR. Environmental impacts of El Arish power plant on the Mediterranean coast of Sinai, Egypt. Environmental Geology. 2002;**42**(6):604-611. DOI: 10.1007/s00254-002-0563-6

[22] Azab MA, Noor AM. Change detection of the North Sinai Coast by using remote sensing and geographic information system. In: The 4th International Conference and Exhibition for Environmental Technologies Environment. 2003

[23] El Banna MM, Hereher ME. Detecting temporal shoreline changes and erosion/accretion rates, using remote sensing, and their associated sediment characteristics along the coast of North Sinai, Egypt. Environmental Geology. 2009;**58**(7): 1419-1427

[24] El-Asmar HM, Hereher ME. Change detection of the coastal zone east of the Nile Delta using remote sensing. Environment and Earth Science. May 2010;**62**(4):769-777. DOI: 10.1007/s12665-010-0564-9

[25] Hamed AAAB. Atmospheric circulation features over the southeasterly part of the Mediterranean Sea in relation with weather conditions and wind waves at Alex. 1979

[26] Nafaa MG, Fanos AM, Elganainy MA. Characteristics of waves off the Mediterranean Coast of Egypt. Journal of Coastal Research. 1991;**7**(3): 665-676

[27] Frihy O, Dewidar K. Patterns of erosion/sedimentation, heavy mineral concentration, and grain size to interpret boundaries of littoral sub-cells of the Nile Delta, Egypt. Marine Geology. 2003;**199**(1–2):27-43. DOI: 10.1016/S0025-3227(03)00145-2

[28] Abo Zed AI, Gewilli YM. Wave climate and coastal response at Rosetta coast, Nile Delta, Egypt. In: Proceedings of 3rd Regional Conference on Arab Water. 2006

[29] Nassar K, Fath H, Mahmod WE, Masria A, Nadaoka K, Negm A. Automatic detection of shoreline change: Case of North Sinai coast, Egypt. Journal of Coastal Conservation. 2018;**22**(6): 1057-1083. DOI: 10.1007/s11852-018-0613-1

[30] Frihy O, Badr A, Selim M, Sayed EW. Environmental impacts of El Arish power plant on the Mediterranean coast of Sinai, Egypt. Environmental Geology. 2002;**42**(6):604-611

[31] Bustillos LV. Gap fill for Landsat 7 images—A correction of SLC-off. 2012

[32] Maanan M, Ruiz-Fernandez AC, Maanan M, Fattal P, Zourarah B, Sahabi M. A long-term record of land-use change impacts on sediments in Oualidia lagoon, Morocco. International Journal of Sediment Research. 2014; **29**(1):1-10

[33] Moore LJ. Shoreline mapping techniques. Journal of Coastal Research. 2000:111-124

[34] Maiti S, Bhattacharya A. Shoreline change analysis and its application to prediction: A remote sensing and statistics-based approach. Marine Geology. 2009;**257**(1):11-23

[35] Alesheikh AA, Ghorbanali A, Nouri N. Coastline change detection using remote sensing. International journal of Environmental Science and Technology. 2007;**4**(1):61-66. DOI: 10.1007/BF03325962

[36] Niya AK, Alesheikh AA, Soltanpor M, Kheirkhahzarkesh MM. Shoreline change mapping using remote sensing and GIS-Case Study: Bushehr Province. International Journal of Remote Sensing Applications. 2013;**3**(3): 102-107

[37] Huang C, Wylie B, Yang L, Homer C, Zylstra G. Derivation of a tasselled cap transformation based on Landsat 7 at-satellite reflectance. International Journal of Remote Sensing. 2002;**23**(8):1741-1748

[38] Thieler ER, Himmelstoss EA, Zichichi JL, Miller TL. Digital shoreline analysis system (DSAS) Version 3.0: An ArcGIS extension for calculating shoreline change. United States Geological Survey, Open-File Report. 2005

[39] Dolan R, Fenster MS, Holme SJ. Temporal analysis of shoreline recession and accretion. Journal of Coastal Research. 1991:723-744

[40] Thieler ER, Himmelstoss EA, Zichichi JL, Miller TL. The Digital Shoreline Analysis System (DSAS) version 3.0, an ArcGIS extension for calculating historic shoreline change. US Geological Survey. 2005

[41] Hwang CS, Choi CU, Choi JS. Shoreline changes interpreted from multi-temporal aerial photographs and high-resolution satellite images. A case study in Jinha beach. Korean Journal of Remote Sensing. 2014;**30**(5)

[42] Ayadi K, Boutiba M, Sabatier F, Guettouche MS. Detection and analysis of historical variations in the shoreline, using digital aerial photos, satellite images, and topographic surveys DGPS: Case of the Bejaia bay (East Algeria). Arabian Journal of Geosciences. 2016; **9**(1):26

[43] Natesan U, Parthasarathy A, Vishnunath R, Kumar GEJ, Ferrer VA.

Monitoring longterm shoreline changes along Tamil Nadu, India using geospatial techniques. Aquatic Procedia. 2015;**4**: 325-332

[44] Hegde AV, Akshaya BJ. Shoreline transformation study of Karnataka coast: Geospatial approach. Aquatic Procedia. 2015;**4**(2015):151-156

[45] Thieler ER, Himmelstoss EA, Zichichi JL, Ergul A. The Digital Shoreline Analysis System (DSAS) version 4.0-an ArcGIS extension for calculating shoreline change. 2009

Section 3

GIS Contribution to Ancient Cities Development

Chapter 4

Contribution of Geographic Information Systems to the Development of Ancient Cities

Mustapha Nassir

Abstract

The world has experienced a significant development of information and communication systems, new technologies, and basic infrastructure. The possession of integrated management systems would contribute to the consolidation of the functional performance of heritage buildings. This study focuses on the ancient city of "Taroudant", located in central Morocco. The geographic information system (GIS) has helped draw maps that identified the geographic data of the city. GIS has also identified the population expansion from 1912 to 2006, which was the reason for the deterioration of heritage buildings in the city. Moreover, GIS has contributed to drawing maps that determine the location of the collapsed parts of the walls, along with the location of historical monuments in Taroudant, facilitating touristic visits and the identification of its features in a short period of time without the need for a tour guide. Presently, modern technologies and applications are among the most important elements supporting the successful transformation of traditional heritage buildings into digital monuments.

Keywords: geographic information systems, ancient city, heritage buildings, geographic data, tourism, development, modern technologies

1. Introduction

The second half of the twentieth century witnessed a technological revolution in information and communication technologies (ICT), which contributed to the approximation of distances and the termination of geographical borders, making the world a small village. Modern technologies and media, such as satellites and the Internet, have facilitated the process of obtaining information from anywhere, bypassing the restrictions of time and place, contributing to clear and significant changes in human life.

In this light, ICT has contributed to changing the daily lifestyle of the people in terms of cultural, social, and economic aspects, especially among young people who have a great ability to deal with any new updates that arise in the field of modern technology, as it facilitates their daily affairs, and helps them to complete many tasks in a short period of time.

The use of ICT led to the progress of many countries and the development of their infrastructure. Therefore, in the face of this technological development that the world has witnessed, it was necessary for the science of cartography to keep pace with this progress by relying on modern technology as the main pillar that would contribute to achieving sustainable development for the domain. This chapter addresses the following problem: How can the applied capabilities and technical components of the GIS (geographic information system) contribute to the development of heritage buildings and to achieving sustainable tourism development?

2. The contribution of GIS to the development of historical sites and heritage monuments

2.1 Identification of the geographical location of Taroudant

Geography is considered among the most important auxiliary sciences to history, as there is a close connection between time and place. Earth is the stage on which the facts of history took place and directed human beings. Therefore, it has the same effect on the course of history, depending on the type of human interaction with the environment and their reaction to its circumstances. Hence, it is necessary for the historian to be aware of the geography of the region (s)he wants to study and the geographical conditions and influencing phenomena affecting it. As a result, and given the importance of geography, the theory of explaining the movement of history through geography emerged. Therefore, historians cannot dispense with geographical studies in all its branches, such as economic, political, and human geography.

The map was considered for a long time an effective tool for storing data related to the geographic area. It is a mean that implements a set of measurements to determine the status of a number of geographical phenomena. Also, maps constituted an essential means of communicating ideas, planning, and implementing projects. Drawing the map requires a long time and exhausting effort. In the face of these difficulties, GIS has emerged as a modern technological tool to overcome these intricacies, allowing geography to adopt a new research approach that focuses on automated data analysis as a new pattern in cartography [1]. This approach aims to provide high-quality services and invest in the intelligence of individuals, institutions, and technologies in order to promote the spirit of innovation [2].

The technological revolution led to many studies on the management of historical monuments, manifested in the use of computers and various programs, along with the prevalence of the GIS, which uses a set of data and information. Moreover, this revolution highlighted the spatial dimensions of heritage sites, identifying their characteristics and features, developing them, as well as planning and marketing them.

Automated maps have contributed to locating many historical cities and archeological buildings. The archeologist records spatial information about the archeological site through maps, which are among the most reliable tools in excavations as they reflect the real value of the archeologist in the context of fossils. Through these maps, the archeologist evaluates the hierarchical order in the formation of the sites [3]. Modern archeologists rely on maps, as they place data from a site within the archeological context by mapping the spatial dimensions and stratigraphy of the site.

In the city of Taroudant, as the focus of this study, GIS contributed to drawing a map about the city's strategic geographical location. Taroudant is located in the center of the Kingdom of Morocco, 80 KMs from the city of Agadir, between longitudes

8° and 49 minutes west of Greenwich, and between latitudes of 30° and 30 minutes north of the equator. The city is located in the middle of a triangular plain (Sous plain) whose summit narrows to the northeast to form a separator between the two chains of the Anti-Atlas in the south and southeast, and the High Atlas in the north. Then, this triangle expands to open to the Atlantic Ocean in the west. Thanks to its positioning, the city of Taroudant and its surroundings are centered in a semi-closed field in the form of a ravine. The city is a lowland confined between the southern margins of the High Atlas and the northern end of the Anti-Atlas. The city is penetrated from the south by valley Souss and from the north valley El Wear [4].

Mapping also contributed to determining the terrain of the region, as it is characterized by the presence of the High Atlas chain, and this is the result of violent elevation movements that caused the formation of the heights of the High Atlas, and this chain is connected to the Small Atlas at the volcanic complex of Sirwa. The terrain takes a hilly character, which becomes more complex when connecting to the Souss plain. Furthermore, starting from the inner Argana lowland, the terrain becomes agglomerated and takes on a mountainous shape that gradually increases in height to reach 4165 m at Toubkal, which is one of the peaks of the mountains N'dern. It is considered the highest summit in this chain and the second highest in Africa [5].

As for the region, there are the Anti-Atlas Mountains, which are mountain blocks with low altitudes that do not exceed 1100 m near the foot of the Sous plain, taking the form of a flat hilly belt in the middle of this chain, as their height increases in the east direction to reach 2531 m (Aklim peaks). The Anti-Atlas is a structural unit with an ancient base that dates back to the pre-Cambrian era and is composed of subterranean rocks covered by primary limestone and schist rocks that take a sloping structural position, allowing them to suddenly sink under the formations of the Sous and Massa plains [6].

Between these two mountains (the High Atlas and the Anti-Atlas) is the Souss plain, which spans an area of 16,100 km2 and opens to the Atlantic Ocean from the west. Its average height does not exceed 300 m, with a general slope west-wise. Regarding the structure of the Souss plain, it can be traced back to the interaction between the Anti-Atlas and the structure of the Atlantic field, ending with the creation of a multi-parameter cumulative structural unit (**Figures 1** and **2**) [7].

These maps enabled us to identify the strategic location of the city of Taroudant, where the geographical and nature conditions of the city contributed to determining its history. Its location behind the Atlas Mountains, which is a natural defensive barrier, made it among the influential cities that control the affairs of the region.

2.2 Identification of the most important monuments in the city of Taroudant

The city of Taroudant is one of the ancient Moroccan cities, which has known many developments throughout ancient historical times in numerous economic, social, and political fields. It is one of the most important cultural and scientific centers in the Souss region, and is considered a link between northern and southern Morocco. Furthermore, Taroudant, being a living proof of the grandeur of the history of this city and its majestic monuments, bears witness to the ingenuity of the craftsmen who constructed its buildings.

We have adopted the GIS framework in drawing a map that pins down the most important military landmarks in the city, such as the historical wall, which is about 7.5 km long, and its height ranges between 7 m and 8 m at the level of the wall parts, along with about 10 m to 12 m at the level of the towers, while its thickness is between 50 cm and 1 m [8]. The Taroudant wall was supported by a dense group of towers

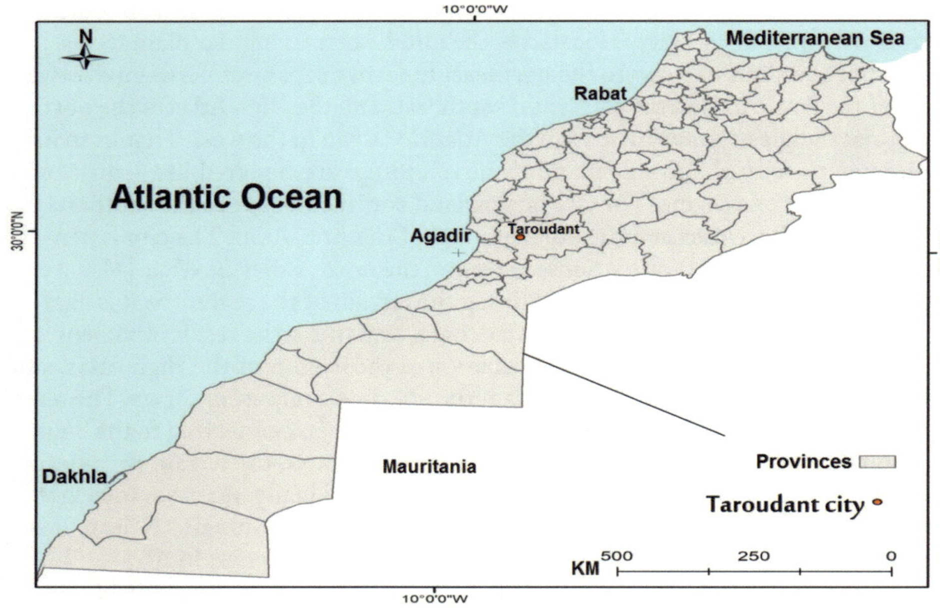

Figure 1.
The geographical location of the study framework (Taroudant city) within Moroccan territory.

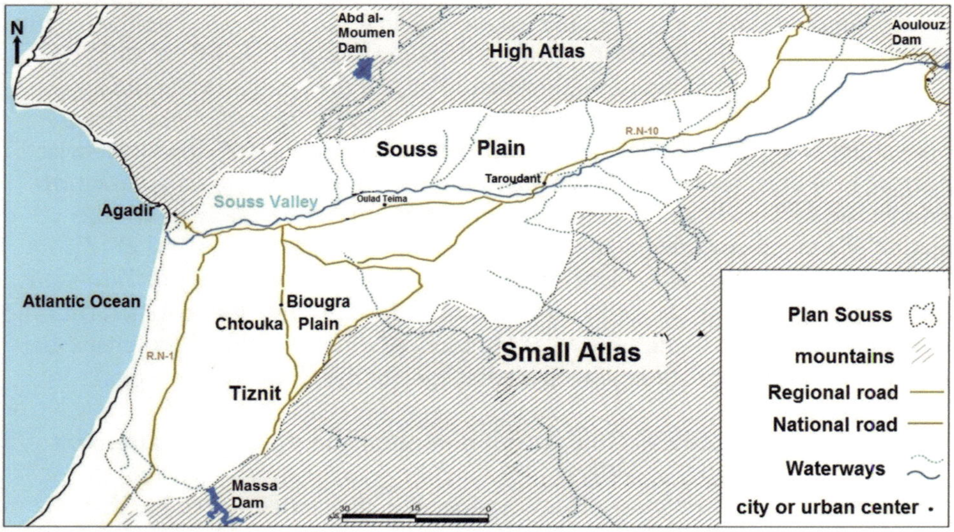

Figure 2.
The location of Taroudant within the Souss plain.

of about two hundred and thirty towers. They are diverse, and some of them are rectangular in shape and numbering about one hundred and sixty-five towers, and others are square and hollow (empty) from the inside, amounting to sixty-five towers, most of which are located on the side next to the kasbah. In addition, the height of the towers of Taroudant city varies between 8 m and 12 m, while the width of the square towers is 5 m, and the width of the rectangular ones is about 10.50 m. The distance between each tower varies according to the walls and is between 6 m and 20 m [9].

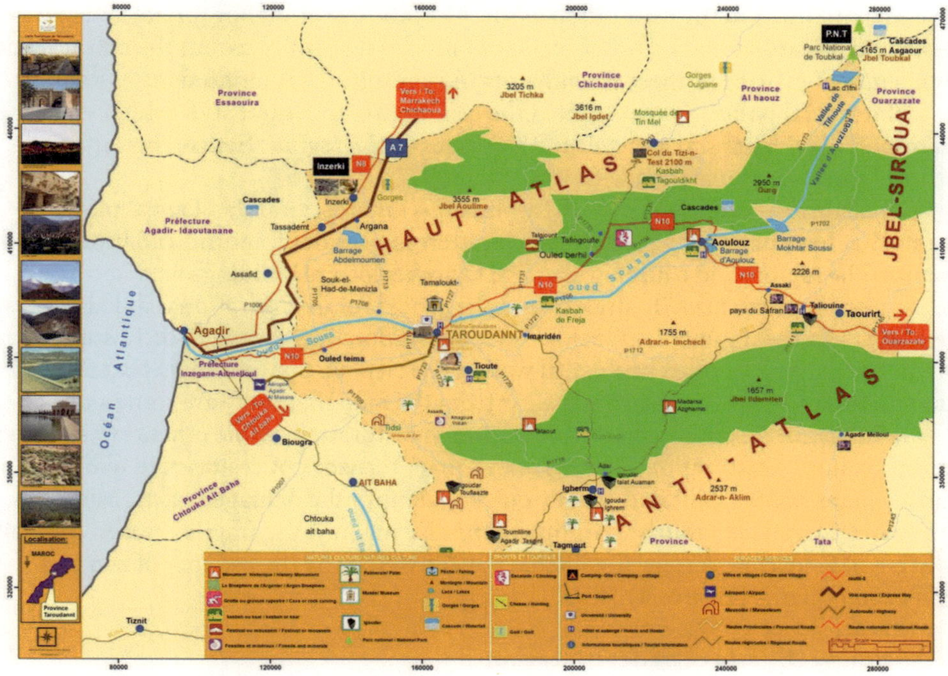

Figure 3.
The most important archeological sites in Taroudant.

The city walls of Taroudant are interspersed with five historical gates surrounding its four sides. The gates were constantly and heavily guarded by watchmen assigned to monitor the arrivals to the city. These gates were closed in the evening and only opened in the morning [10]. These gates retained the same names they were known by. However, the shape of their architecture differs from one gate to another. The five gates are: Bab LKhamis, Bab Targount, Bab Zourghan, Bab Ouled Bounouna, and Bab Salsla.

A number of civil landmarks, such as Riads, hotels, traditional baths, and waterways, were also localized on the map. Also, many historical religious monuments such as mosques, shrines, and marabouts that represent the icon of Islamic civilization are identified on the map. This invaluable heritage is still vibrant is observed in the most important buildings built in the city of Taroudant thanks to the map that helped facilitate access to these monuments, especially since the city's road network is intertwined and complex (**Figure 3**) [11].

3. The role of GIS in diagnosing the engineering condition of the heritage buildings in Taroudant and the reasons for their deterioration

3.1 Diagnosing the engineering condition of the heritage buildings in Taroudant

Most of the heritage urban monuments scattered in the ancient city of Taroudant suffer from destruction and deterioration, as many of the heritage monuments in the city have perished as a result of neglect. The rest of them have been subjected to a major distortion in their architectural features, due to the use of modern materials (cement and iron) in the restoration work. Also, the use of inappropriate and extraneous modifications and techniques did not mostly take into account the cultural and civilizational privacy of these monuments. As a result, these monuments witnessed an architectural pollution that spoiled the original architectural style which was based on respecting the natural and cultural privacies of the region. Hence, this authentic cultural capital is threatened with vanishment [12].

The historical wall of the city of Taroudant is among the archeological monuments that have been subjected to deterioration as a result of natural and human factors. The factors and causes of damage to archeological buildings vary according to numerous circumstances. To understand the process of deterioration that led to these problems, the most important factors of deterioration must be studied. To study the current situation of the archeological wall of the city, a map was drawn that showed its architectural condition, and it is clear from the map that there are some sides of the wall that are deteriorated and not subject to restoration, while others are in better condition due to restoration. Thanks to this map, the work of restorers in maintenance operations is facilitated as they quickly identify the damaged parts. Moreover, this map will also help tourists coming to the city to know the fragile areas in the wall, which will help avoid any accidents that may be caused by the collapse of some of the wall's components (**Figure 4**) [13].

3.2 Diagnose the causes of deterioration

Among the factors that lead to the deterioration of the heritage monuments in Taroudant are the polluting activities that excrete several gases such as sulfur dioxide, carbon dioxide, and nitrogen compounds that turn into acids when certain conditions are present, leading to changes in the color of the walls' coating, in the form of black layers. These factors resulted in the decomposition and loss of components of the walls' building materials. Regarding the building materials that are most susceptible to damage by acid pollution gases in historical buildings, we recognize gypsum and limestone, in addition to some sandstone and marble [14].

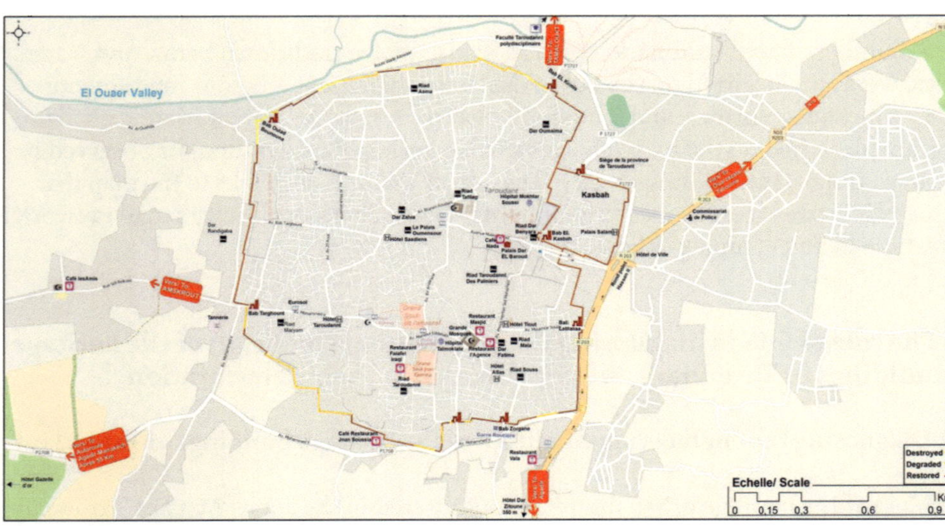

Figure 4.
The current engineering condition of the wall.

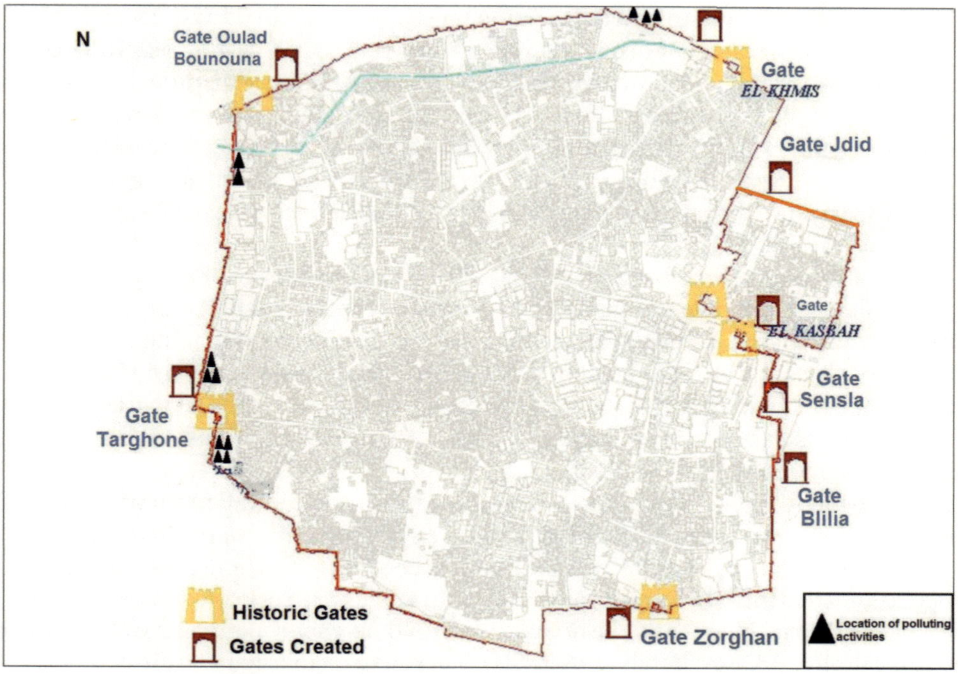

Figure 5.
Shops of polluted activities in the ancient city of Taroudant.

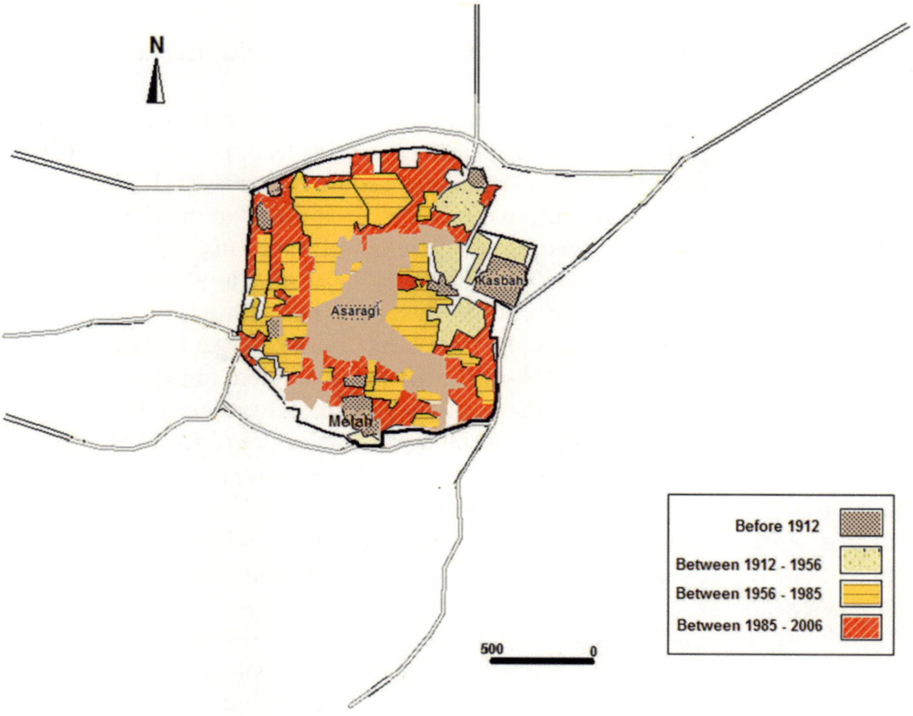

Figure 6.
Urban expansion in the ancient city of Taroudant from 1912 to 2006.

The GIS contributed to drawing a map that showed the spread of craft activities that harm the environment inside the ancient city of Taroudant. It is clear from the map that most of the contaminated shops are located near some historical gates, specifically at the historic city (Taraghont and Zorghan, Lkhamis), and in separate ancient neighborhoods in the city. Furthermore, some black spots of grease as well as paint on the sides of the wall are noticed, as craftsmen place and paint metal pieces on the walls, causing distortion of the urban view of the archeological wall surrounding the city of Taroudant [15].

Also, among other main reasons that led to the disappearance of many monuments in Taroudant is the urban expansion, and it is clear from the map that the ancient city knew during the colonial period a slight urban development that included at first its eastern side and the roads leading to the Kasbah Gate. However, after Morocco gained independence in 1956, the urban expansion began to rise exponentially, as new neighborhoods have appeared, centered mainly around the center of Asarag Square and the northwest side of the old city.

The residential blocks began to expand near the historical wall in a horizontal and vertical manner, and the reason for this urban expansion is due to the high population growth in the city (838,820 people) [16], according to the figures of the General Population and Housing Census for the year 2014. This expansion led to the creation of new residential blocks and the demolition of old neighborhoods and archeological buildings (olive presses, hotels, and Riads). These expansions led to a distortion of the general landscape, due to the construction of modern buildings at high altitudes, which contributed to visual pollution in the areas and neighborhoods in the ancient city (**Figures 5** and **6**) [17].

4. GIS and its contribution to sustainable tourism development

In recent years, global attention has focused on cultural tourism, due to the great economic value that it plays in providing jobs and bringing in important revenues as well as being the main source of national income in some countries [18]. Museums, historical sites, and products of traditional industry, in addition to the elements of intangible cultural heritage consisting of customs and traditions, represent an essential component of tourism activity and cultural heritage. They provide multiple and diverse resources that constitute basic products that fall within the tourism offer. According to statistics conducted by the World Tourism Organization in 2007, cultural tourism constituted about 40% of the total international tourism [19].

Heritage sites are one of the most important aspects of cultural tourism, and attract tourists from different countries of the world. In the face of the increasing demand for cultural tourism, traditional heritage buildings have become unable to provide good services to tourists, which has led to a decline in their competitiveness. Nevertheless, advanced means and applications, such as GPS technology, mobile communication system (5G–4G-3G-UMTS), and automatic maps, will help improve tourism services [20].

These advanced means facilitate touristic visits to heritage sites along with short quick identification of monuments without the need for a tour guide. They can also download in the Play Store or Apple Store an automated map that contains basic information about the site and heritage buildings located in the city, which will enable tourists to access these historical sites easily and in a safe, uncrowded road, at low cost and in a short period of time. Automated maps allow the identification of dangerous heritage sites that are at risk of collapsing, as well as those that are difficult to reach or that have been destroyed. The application of these technologies will enable the provision of an advanced and sustainable tourism product, based on flexible knowledge bases. In addition to that, it will provide high-quality services to tourists with minimal effort and cost, which will raise the level of comfort and luxury for all, and will positively affect the movement of investment and the economy [21].

5. Conclusion

The GIS contributed to drawing accurate automated maps that identified the strategic location of many historical cities, and this system also helped to localize the most important military, religious, and civil monuments in the ancient city of Taroudant on automated maps that have contributed to accessing these historical monuments easily especially in light of a complex road network. In the city, GIS also helped in diagnosing the engineering condition of the heritage buildings, where a map was drawn showing the architectural condition of the historical wall of Taroudant, which will facilitate the work of restorers in maintenance operations, by quickly identifying the damaged and vulnerable parts.

The GIS also contributed to the development of a map of the settlement of craftsmanship activities that pollute the environment, which caused the deterioration of some sides of the walls and distorted its urban landscape and the city's monuments. The automated maps that can be loaded into Android system applications will help tourists to visit archeological sites easily, and avoid dangerous areas that may lead to dangerous accidents that threaten the safety of tourists.

Acknowledgements

Many thanks to Dr. Ahmed Chaib for supervising this work, Dr. Miloud Talibi (GIS Specialist) for designing the maps used in this work, Nourddine Binij (researcher) for correcting and proof-reading, and the Regional Delegation for Tourism in Taroudant for providing maps.

Author details

Mustapha Nassir
Faculty of Letters and Humanities, Ibn Zohr University, Agadir, Morocco

*Address all correspondence to: mustaphanasir1981@gmail.com

References

[1] Chaaouan J, Faleh A. Geographical Information Systems and Remote Sensing Principles and Applications. Info: Print; 2012. pp. 13-14

[2] Houlin Z. Accelerate the development of smart and sustainable cities, build the smart and sustainable cities of tomorrow. Union News Journal. 2016;**02**:1

[3] Méthodes d'excavation [Internet]. Available from: https://www.encyclopedia.com/science/encyclopedias-almanacs-transcripts-and-maps/excavation-methods. [Accessed: 2022-22-02]

[4] Khalil S. Environmental reflections of certain land movements in Taroudant. In: Benhalima H, Coordinator, editors. Taroudant, the Capital of Souss. 2nd ed. Rabat: Maarif El jadida; 1997. p. 230

[5] Ait Hsaine A. The alluvial deposits of the Taroudant site and their geomorphological significance. In: Benhalima H, coordinator, editors. Taroudant, the Capital of Souss. 2nd ed. Rabat: Maarif El jadida; 1997. p. 39

[6] Bouchelkha M. The Souss Massa Countryside, Modern Transformations and Socio-Spatial Dynamics. 1st ed. Marrakech: Library al watania; 2007. p. 39

[7] El Mahdad H. Elements of the natural environment of the Souss. In: Benhalima H, coordinator, editors. Taroudant, the Capital of Souss. 2nd ed. Rabat: Maarif El jadida; 1997. p. 211

[8] Abu Rehab M. The Islamic Monuments of the Far Maghreb in the Merinid and Saadian Periods. 1st ed. Alexandria: Dar Alwafae; 2015. p. 157

[9] Field work of the researcher

[10] De Torres D. Chorafa History. Casablanca: Publishing and Distribution Company; 1988. p. 34

[11] Field work of the researcher

[12] Field work of the researcher

[13] Almahari S. Conservation of Historic Buildings in the City of Muharraq. Regional Center for the Preservation of Cultural Heritage in the Arab World. Sharjah: Sickroom; 2017. p. 129

[14] Field work of the researcher

[15] Field work of the researcher

[16] High Commission for Planning. Regional Statistical Bulletin Souss-Massa. Agadir: Regional Planning Delegation; 2017. p. 15

[17] Field work of the researcher

[18] Joëlle S. Ville Intelligente Comme Vecteur Pour Le Développement Durable Le Cas De La Ville Montréal. Sherbrooke: University; 2015. p. 13

[19] Centre for Entrepreneur Ship, SMES, Local Development (OECD). The Impact of Culture on Tourism. Paris; 2009. p. 18

[20] Sandra B, Jérémy D. La Ville Intelligente Origine. Définitions, Forces Et Limites d'une Expression Polysémique. Québec: National Institute for Scientific Research Center Urbanization Culture Society, Montréal; 2017. p. 15

[21] Nikos K. Intelligence Cities and Globalization of Innovation Networks. Casablanca: Routledge; 2008. p. 8

Section 4

GIS Applications in Agriculture and Land Suitability Analysis

Chapter 5

GIS Applications in Agriculture

Parmita Ghosh and Siva P. Kumpatla

Abstract

Technological innovations during the recent centuries have enabled us to significantly boost agricultural production to feed the rapidly increasing global population. While advances in digital technologies triggered the onset of the fourth revolution in agriculture, we also have several challenges such as limited cropland, diminishing water resources, and climate change, underscoring the need for unprecedented measures to achieve agricultural resilience to support the world population. Geographic information system (GIS), along with other partner technologies such as remote sensing, global positioning system, artificial intelligence, computational systems, and data analytics, has been playing a pivotal role in monitoring crops and in implementing optimal and targeted management practices towards improving crop productivity. Here we have reviewed the diverse applications of GIS in agriculture that cover the entire pipeline from land-use planning to crop-soil-yield monitoring to post-harvest operations. GIS, in combination with digital technologies and through new and emerging areas of applications, is enabling the realization of precision farming and sustainable food production goals.

Keywords: Geographic Information system (GIS), precision agriculture, remote sensing, Global Positioning System (GPS), resource management

1. Introduction

As the world population is projected to grow close to 10 billion by 2050, we need to produce about 50% more food compared to 2013 production to meet the global demand [1]. This goal needs to be met while facing the challenges of climate change, the limited scope of arable land expansion, and dwindling water resources. In addition, anticipated food production also needs to incorporate practices for sustainable management of croplands to preserve soil health, conserve water resources, and encompass biodiversity [2]. Considering these challenges and constraints in achieving our food production targets there is an unprecedented need for monitoring of crop growth and health and timely interventions to maintain or improve crop productivity while reducing wastage of inputs and resources. Advances in sensors, communication technologies, computational systems, and powerful data analytics are enabling us to accomplish these tasks. Technologies that can enable efficient use of agricultural inputs and reduce environmental losses while contributing to increased and sustainable production are of great value for achieving food security. Several existing and emerging tools and technologies such as geographic information system (GIS), remote sensing (RS), Global Positioning System (GPS), Artificial Intelligence (AI), Big Data

Analytics, and Internet of Things (IoT) are instrumental in achieving this goal through efficient monitoring of crops and soils, and, combined with other pieces of information, are providing data-driven insights for targeted or site-specific management of crops ensuring increased productivity [3]. Geographic Information System (GIS), a key foundational technology, is defined as a powerful system comprising tools for the collection, storage, and retrieval of data at will, as well as analyzing, transforming, and displaying the spatial data for a specific purpose [4–7]. It plays a critical role as it provides the spatial context and information on several features each of which is available as a data layer. In addition, it provides the tools to manipulate spatial and non-spatial data and presents them through intuitive and illustrative map formats [8]. GIS has been making an impact in diverse domains that include geography, environmental sciences, natural resources, forestry, agriculture, food, manufacturing, banking, and health services [8]. Recent decades have seen a significant increase in the application of GIS tools for diverse applications in agriculture at local, regional, national, or global scales. These applications most often involve the use of GIS along with partner technologies such as remote sensing, GPS, and data analytics towards an in-depth understanding of a given farm or a region and facilitating intervention or corrective measures for the crops and/or the soils. Since the GIS data are linked to a common referencing system, another advantage GIS offers is that the same data can be used for different applications or goals and we can also bring in other data and, combining that with existing data, we can perform a joint analysis for deriving novel insights. Many studies have reported the use of GIS for diverse applications in different crops [9–11]. To further enable the readers to develop a strong appreciation for the role of this powerful technology in agriculture, here we have reviewed the most widely used and emerging applications of GIS, either by itself or in combination with other partner technologies, and how it has been making major impacts on agricultural productivity and supply chains.

2. Key partner technologies of GIS

The power and impactful contributions of GIS in diverse domains can be attributed to the combined use of GIS and two other key geospatial technologies: GPS and Remote Sensing. Each of these three partner technologies plays a crucial role in realizing the goals of applications (Section 3), and, therefore, are briefly described below.

2.1 Geographic information system (GIS)

Based on its role in supporting the collection, storage, retrieval, and analysis of data on features and location, and its utility for data-driven solutions, especially in site-specific management, GIS is considered the brain of Precision Agriculture [12]. Digital GIS maps differ from conventional maps in that they harbor several layers of information each layer providing information or a map about a given attribute such as soil survey, precipitation, nutrient status, pest infestation, yield, etc. In addition, GIS provides the analytical capability by using statistical tools and geospatial analytics enabling extraction of inter-relationships between attributes, and the insights, thus derived, are valuable for decision making with respect to management practices.

2.2 Global positioning system (GPS)

This positioning/navigation system based on a satellite network enables the determination of positional information by providing the latitude, longitude, and elevation of a location. The location information collected by GPS receivers enables farmers and researchers to the reliable identification of fields, mapping of field boundaries, water bodies, infested or problematic areas in the field, and for understanding the relation to several other attributes within and outside the boundaries of a given field. Such a high-fidelity field mapping permits site-specific application of nutrients, pesticides, herbicides, and water, thereby improving productivity and reducing input costs—the essence of precision agriculture.

2.3 Remote sensing (RS)

Remote sensing, with its diverse methods and applications in agriculture, has revolutionized crop monitoring and interventions for improving farm productivity [13, 14]. RS, in combination with GPS, GIS, and other tools is critical for implementing the goals of precision agriculture. This combination is crucial for enabling several applications that provide the basis for site-specific management of fields and include soil mapping, crop growth monitoring, estimation of soil moisture and fertility, detection of biotic (pests and diseases) and abiotic (drought and flood) stresses, and yield estimation.

3. GIS applications

The onset of digital agriculture, considered the fourth revolution in agriculture, has totally transformed the way farming is done, thanks to advances in geospatial technologies, sensors, artificial intelligence, robotics, and other tools and technologies. The ability to precisely identify the problem areas in cropland and monitoring and management of all steps in the entire agriculture value chain requires image and non-image data along with spatial context. GIS, with its component tools and analytic modules, and the data gathered by its partner technologies like remote sensing and GPS provides intuitive and lucid visualization of information for data-driven decision making for improving crop productivity. While GIS has been used for agricultural applications for quite some time, the number of applications has been growing rapidly in recent years due to technological advances. Several most common and emerging applications are presented and discussed below.

3.1 Land suitability assessment and land use planning

We are in an era where we are facing the challenge of feeding billions of people while the fertile land is shrinking, therefore, we need to optimize the use of natural resources to maximize the benefits. GIS provides an excellent platform for assessing the quality of land for suitable applications. Multi-criteria decision-making (MCDM) approach based on GIS is the most popular choice among researchers for land use planning. Researchers use different features offered by GIS such as soil type distribution, soil texture map, buried deep underground water level distribution, soil fertility distribution, soil pollution distribution, hydraulic conductivity of soil (Ks), slope (S), soil texture (ST), depth to water-table (DTW), and electrical conductivity of groundwater (ECw), climate conditions, topography, and satellite data, and identify the variety of interactions, dependencies, and the impact of these interacting factors on sustainable land use.

Chen et al. [15] evaluated weight sensitivity of MCDM model for land suitability assessment for irrigated agriculture. They aimed to examine the sensitivity of changing weights of the input features on the model output. The results suggested a strong influence of sensitivity and, therefore, they have recommended giving special emphasis on this criterion. Zolekar and Bhagat [16] have used GIS-based MCDM model with IRS P6 LISS-IV images as input for the evaluation of agricultural practices in hilly regions. The rank of influential criteria was determined by correlation analysis and recommendations from scientific literature. The combined use of remote sensing and GIS turned out to be beneficial for land suitability evaluation. Pan and Pan [17] applied three scales, two-step analytic hierarchy processes (AHP) for GIS-based crop suitability assessment. They have emphasized the importance of selecting appropriate evaluation factors and suggested the consideration of features with a significant difference and controlling the land use and avoiding causality. Following this approach of feature selection, the AHP output was spatially distinct. The authors have recommended appropriate land use based on land suitability maps. In another study, [18] selected the features based on growth requirements for examining the land suitability for the wheat crop. Analytic Network Process (ANP) model was deployed for assessing the interdependence of strategic input features for site suitability evaluation of citrus crops [19]. The ANP coupled with GIS–MCDM identified critical factors for maximizing yield and minimizing production loss. AHP integrated with geo-statistics had proven its merit for maize cultivation land suitability mapping in calcareous and saline-sodic soils [20]. These powerful GIS tools enable land reclamation planning with suitable conservation practices.

Integrated fuzzy membership and GIS model were used to analyze arable land suitable for farming. Topography and eight soil parameters were utilized for fuzzy membership classification and the important crop productivity-related soil features were accommodated accordingly. Fuzzy membership allowed the consideration of partial memberships which is unlikely in classical approaches for classification. This self-adaptive approach revealed that the land was better suitable for groundnut cultivation contrary to the current practice of Finger millet cultivation. Results of this experiment proved that the GIS-based decision system can surpass the traditional knowledge and, if deployed accurately, can improve the productivity of land [21]. This is the need of the hour technology as land and natural resources are declining, and the demand for food production is increasing rapidly. The fuzzy set model, AHP, and GIS were combined to generate a land suitability map for tobacco production [22]. This study has once again demonstrated the advantage of using Fuzzy membership functions for land suitability analysis. AHP has the power of accurately assigning weights to the input factors in a logical way. The maps were generated by ArcMap. The integrated application of fuzzy, AHP, and GIS helped to circumvent the problems resulting from the uncertainties, subjectivities, and hierarchy characteristics of the traditional land suitability assessment process. GIS is a powerful tool to delineate the study area, manipulate geographic data, process maps, and present results in land suitability assessment. Integration of Fuzzy set and AHP methods with GIS provides a precise and powerful combination in applying for land suitability analysis. Researchers advocate that Fuzzy logic coupled with other decision-making methods is one of the best approaches for land suitability analysis [21–23]. Scientists are also exploring artificial intelligence along with GIS for efficient land use planning [23].

3.2 Water resource management

Abundance of water supply is a primary requirement for meeting the demand for food production by the ever-increasing global population. As indicated earlier, farmers have the responsibility of feeding about 10 billion people in 2050 which demands a 50% increment in food production compared to 2013 level [24, 25]. The availability of clean water is decreasing and dependence solely on rainfall is not a viable option for the farmers [26, 27]. In this challenging scenario, water resource management is the key to success. Irrigation is the best solution for meeting the water requirement in agriculture. GIS technology backed by remote sensing has already proved its merit for the management of water resources [28–31]. Researchers strongly suggested that remote sensing can supplement the traditional geophysical models for groundwater potential assessment and recharge experiments [32, 33]. Many researchers supported the potential of GIS for groundwater management [34, 35]. Tripathi et al. [36] integrated the MODFLOW groundwater model with the GIS for watershed prioritization. Singh et al. [37–39] combined GIS and remote sensing for delineating groundwater potential zones. Lineament and hydro- geomorphological maps were prepared from remote sensing images. The delineated groundwater potential zones have been found to show synergy with the well-yield data. When sub-watershed level runoff and sediment yield were assessed using the combination of GIS and remote sensing data it reduced the time of the input data process and produced good results compared to actual runoff and sediment yield [40]. Determining the suitability of irrigation for a given geography is one of the most popular applications of GIS. A study conducted in UAE accounted for non-renewable sources like desalination and treated sewage effluent (TSE) to assess irrigation suitability [41]. This type of water-scarce region needs optimization of water resources management. Land management, topography, climate conditions, soil capabilities, and water potential were used in the analytical hierarchical process (AHP) GIS model to assess crop suitability. The results showed that the land was unsuitable for cereals and vegetables but the cultivation of sorghum, jojoba, fruits, date palm, and forage was recommended. This study unleashed the power of GIS technology for using every acre of fertile land in a geography with a high level of water scarcity. Reduction in clean water resources is motivating researchers and policymakers to identify suitable alternatives for irrigation water. Zolfaghary et al. [42] evaluated the scope of using urban treated wastewater as an alternate source of irrigation. They have utilized the MCDM method which was executed in the GIS software environment and the Analytic hierarchy process (AHP) was used. This analysis revealed that the suitability of treated wastewater is subject to suitability for crop cultivation, nitrate contamination burden, and aquifer vulnerability. ISAREG irrigation scheduling model was integrated with GIS with the aim of generating efficient irrigation scheduling advice and identification of practices to account for water savings and salinity control [43]. These results advocated the successful outcome of the model for irrigation scheduling and choosing water saving measures during both wet and dry years. Though the intensification of irrigation is beneficial for food production, it was pointed out that soil salinization and waterlogging are the major drawbacks of irrigated agriculture intensification, and that strong emphasis should be put on leveraging GIS and remote sensing technology for monitoring the problem areas followed by planning conservation and preventive measures [44].

3.3 Soil health and fertility management

Soil fertility is directly proportional to productivity. It controls the availability of nutrients and water to the crop. The soil fertility has been degrading due to various factors like pollution, sealing, overgrazing, waterlogging, excessive use of agricultural chemicals, and erosion. It is crucial to determine soil health and fertility status for planning effective practices for site-specific management or precision farming [45–47]. Soil macronutrients (N, P, and K), micronutrients (Zn, Mn, and Fe), pH, soil organic carbon (SOC), water holding capacity, erosion status, and moisture content are extensively used features for soil fertility status assessment [48–51]. Spatial interpolation, Multi-Criteria Decision Analysis (MCDA) [52–55], and Ordered Weighted Averaging (OWA) [56–59] are the most popular geospatial analysis techniques which provide spatiotemporal variability of soil health and fertility status to the decision-makers.

Soil erosion status is an essential parameter for soil quality assessment and spatial variation in erosion gives a clear picture for agricultural planning [60]. It was demonstrated that geospatial maps of soil erodibility generated by Inverse Distance Weighted (IDW) method is a great tool for assisting in sub-watershed level land use planning. AbdelRahman et al. [61] combined the remote sensing and GIS technology to assess the soil fertility status. They have used the LISS III and IV images for land use classification and the RUSLE method for soil erosion estimation, collected the soil nutrient field data, and applied a geostatistical model to identify the spatial variation of soil erosion and nutrient availability. In another study, [62] used the IDW model for derived soil nutrient maps and applied the OWA method to make the maps homogenized and used those as the input for the fuzzy inference system for soil fertility mapping. Fuzzy mathematics developed with soil organic matter (SOC), total N, total P, total K, available N, available P, available K, pH value, and cation exchange capacity as indicators in ArcGIS showed that the soil fertility of mid- and low- yielding fields were low and are directly correlated with soil profile configuration [63]. The association of crop productivity with the soil fertility is evident and GIS-based soil maps and fertility status give prior information about the field-specific crop suitability.

Leena et al. [64] proposed GIS-enabled cloud technology for soil fertility management decision support system. This system has the capability to make fertilizer recommendation based on soil test and crop response. This recommendation system helps farmers optimize their fertilizer usage and maximize yield. This system generated spatial nutrient variation works as fantastic e-governance system for the government agencies. GPS- and GIS-based soil fertility maps are great tools for thorough monitoring of the soil health and, based on such maps, [65] recommended application of paper mill sludge to reduce acidity in the soil and cultivate pulses and groundnut to make the best use of the acidic soil. These geospatial soil maps have proven to be an effective decision support system in the context of food production challenges due to soil degradation. Tunçay [66] applied soil fertility index (SFI) based on the variables of sand, silt, clay, pH, EC, OM, CaCO3, Ntotal, Pavb, Kexc, Caexc, Naexc, Mgexc, and available micronutrients (Feavb, Cuavb, Znavb, Mnavb) and proved the strength of SFI. This study demonstrated the potential of combining Sentinel 2 image-derived crop yield for validation of soil fertility model. Advances in the observatory systems such as remotely sensed data of fine-to-coarse spatiotemporal resolutions, and in the process-based and data-driven modeling techniques have facilitated the collection, storage, analysis, visualization, and interpretation of non-spatial data for soil fertility index (SFI) [67–72].

Li et al. [72] applied weighted space fuzzy clustering coupled with the soil nutrient space mutation distribution for soil fertility characterization. This information aids in optimizing the fertilizer recommendation system. Agricultural practices such as crop residue management, nutrient management, soil tillage, and pest management affect ecosystem goods and services and soil quality and fertility [73–75]. The best management practices, compatible land use/cover changes, and land suitability analysis are required to prevent the degradation and loss of prime farmlands [73, 76, 77]. Soil erosion management, soil biodiversity improvement, and rehabilitative farming systems are some of the best management practices used to improve soil quality and crop yields [78–80].

A study that leveraged GIS and fuzzy evaluation method to evaluate the soil fertility status used total nitrogen, total phosphorus, total potassium, available nitrogen, available phosphorus, available potassium, soil organic matter, cation exchange capacity, and pH as indicators for the generation of fertility indices [63]. This fertility index revealed that total nitrogen and soil organic matter are higher for paddy fields. These fertility maps also give an insight into the suitable soil qualities under different types of land use and climatic conditions. Sub-watershed level nutrient mapping revealed that available N, P, S, Zn, and Fe are controlling agents of soil fertility [81]. Thus, fertility maps and their relationship with soil properties and crop yields serve as an information system for precision agriculture.

3.4 Biotic and abiotic damage assessment and intervention

Studies have reported that biotic crop damage, caused by insects, fungi, and other pests, can cause 15–70% yield loss [82–84]. This scenario impacts the demand and supply chain and also affects the economy of farmers. The changing pattern of weather makes the crops susceptible to pests and diseases. The availability of crop protection methods is quite beneficial for tackling crop health, but the lack of timely information about the pests and diseases makes the damage irrepressible. GIS technology holds immense potential for site-specific pest and disease management. Remote sensing and GIS-based forewarning systems are boon to farmers to arrest the yield and economic loss. Ranjan and Vinayak [85] advocated that pest and disease forecasting systems allow farmers to apply the control measures in time to reduce the cost of production. Apart from the forewarning system, the pest population density map also plays a crucial role in identifying the hotspots and extending advisory to farmers. According to [86], information about the geospatial density of oriental fruit moth, *Grapholita molesta*, had the scope of reducing the crop injury and pest population by applying geographically suitable management measures. The difference in current and predicted geospatial distribution of two polyphagous and invasive Icerya species clearly indicated the impact of climate change in modifying the pest attack patterns [87]. Such pest distribution maps enable farmers, agricultural experts, and policymakers to prepare management strategies to combat pest attack in the future. Tracking the migratory patterns of pests is of utmost importance given the instances of a sudden outbreaks of cutworm, *Agrotis ipsilon*, and fall armyworm, *Spodoptera frugiperda*, in different continents than their usual geographical locations [88–90]. Remote sensing and GIS are important tools for monitoring habitats of pest species such as western tarnished plant bug, *Lygus hesperus*, and Migratory and Australian Plague locusts [91, 92]. Remote sensing and GIS are rapid and cost-effective technology for assessing the extent of crop damage by pests and diseases [93]. Researchers had demonstrated the feasibility of pest and disease type detection and severity mapping from remote sensing images [94–96]. These damage assessment maps hold key spatial information about the damaged crop acreage and its trend across multiple years over different geographical units. These maps have the potential to act as an aid for insurance settlement for the farmers and for those seeking government subsidies and benefits.

Natural calamities cause irreversible damage to agriculture. Rapid mapping and quantification of damage aid in economic loss recovery and act as a decision support system. A geo-spatial model is used in a case study to assess the impacts of extreme flood events on agricultural production in the Quang Nam province of Vietnam. Chau et al. [97] generated the water surface by interpolating flood depth marks by the inverse distance weighting (IDW) and employed a digital elevation model (DEM) to generate the flood inundation map. This map overlaid with the land use map gave an effective estimate of the damaged agricultural area [98–100]. Drought is another constraint to agricultural productivity and understanding the hotspot and climatology is crucial to strategically minimize the impact. MODIS satellite Normalized Difference Vegetation Index (NDVI) derived drought risk classes were prepared to access the spatial pattern [101]. GIS-based characterization of climate variability and drought zones provides scope for strategic measures adoption to maximize productivity [102, 103].

3.5 Crop monitoring and yield prediction

Monitoring of crop growth, health, and accurate or near accurate prediction of yield is crucial not only for estimating economic return but also for assessing the food production thereby helping in the management of food security. Many studies showed that traditional methods of crop yield estimation could lead to poor assessment and inaccurate crop area appraisal [104, 105]. Moreover, these methods require time-consuming, labor-intensive, and expensive crop and yield data collection. This is where technologies like remote sensing (RS), GPS, and GIS provide a huge advantage as they can be used to assess temporal and spatial variability of crop dynamics and yield output [106]. The use of two key partner technologies, RS and GIS, with required input from others can provide an efficient solution for monitoring crop health and developing models for predicting crop yields across diverse spatial scales. While remotely sensed images and associated analytics permit the tracking of crop health and predicting the yield, GIS technology enables the collection, storage, retrieval, and visualization of data that were linked geographically. Remotely sensed geospatial data acquired by satellites, aircrafts, or unmanned aerial vehicles (UAVs) can be used to gather information on several features of the crops and the characteristics of the soils supporting their growth thereby enabling the assessment of crop health. The images gathered can be used for assessing general vigor, disease or pest infestations, or deviations from expected growth due to drought or other abiotic stresses. Geospatial data collected in a spatiotemporal manner and the associated analysis techniques help in assessing the changes in the health of crops thereby permitting management interventions while providing predictions on anticipated yields based on the growth and health of the crops. A commonly used method for assessing crop health is based on the determination of vegetation indices that are calculated based on surface reflectance from crop canopies at two or more wavelengths. Many vegetation indices are available for evaluating the extent and vigor of vegetation, crop growth dynamics, stress due to biotic or abiotic factors, and other useful assessments

[107]. Adhav et al. [108] used multiple vegetation indices that included Normalized Difference Vegetation Index (NDVI), Green Normalized Difference Vegetation Index (GNDVI), Chlorophyll Vegetation Index (CVI), and Difference Vegetation Index (DVI) to determine crop health as well as variations in health conditions. To further improve the efficiency of health assessment, they have combined all vegetation indices using ArcMap 10.5 software and reclassified the merged indices which were then used for categorical representation of health scenarios. Such a representation helps farmers to identify areas that need immediate management intervention [108]. Determination of crop health is particularly critical in smallholder farms as the subsistence and livelihood of these farmers depend on the productivity of their crops. As per a recent study, small farms were found to account for 84% of all farms worldwide but they operate only on around 12% of all agricultural land and produce about 35% of world's food [109]. Use of UAVs for gathering and leveraging data for assessing crop growth and dynamics has proven to be crucial for farmers to take timely and appropriate corrective measures to maintain or increase productivity. A recent study in South Africa [110] evaluated the utility of multispectral UAV imagery and random forest machine learning (ML) algorithm to estimate maize chlorophyll content at various growth stages and created a chlorophyll variation map capturing the spatial heterogeneity of chlorophyll in the field thereby helping the farmers to take management actions. Considering the strategic role of sustainable intensification towards the food production goals of Sub-Saharan Africa [111], such RS- and GIS-based diagnostics and interventions are critical for smallholder farmers. Another study assessed crop health using different chlorophyll indices in addition to modified vegetation index by leveraging data from two different satellites and ArcMap (of ArcGIS) for geospatial analytics [112] resulting in insights that could be used for managing nutrient applications towards improving crop productivity. Two essential prerequisites to implement location-specific management practices and interventions are the availability of an accurate acreage map of crop of interest and the cropping systems of a given area and technologies for predicting yield before the reproductive phase or harvesting of the crop. The use of RS and GIS technologies can help achieve both goals. NDVI, a commonly used vegetation index, serves the dual purpose of assessing crop health and predicting crop yield while GIS tools can provide the spatial context. Several studies abound that leveraged NDVI and GIS for yield predictions and a few examples are discussed here. In a study that measured NDVI values at different growth stages of rice, several linear regression-based yield prediction models were developed using NDVI values and narrowed down to a model that had the highest prediction potential and was also able to predict yield well ahead of harvesting time [113]. Such a model can help the farmers to implement changes to the fertilization, water, pest, and disease management practices towards realizing improved productivity. Using time-series data of SPOT vegetation and two key spectro-agrometeorological variables, rainfall estimate (RFE) and NDVIactual (NDVIa), that are highly correlated to maize yield, [114] have developed an operational model with high predictive ability for yield forecasting in Ethiopia. By leveraging both RS and GIS, this model enabled yield forecast at flowering season which is more than two months earlier than the forecast by conventional method thus providing an advantage of early intervention towards crop productivity and crucial data for the authorities for crop production estimates [113]. In a field study aimed at developing an efficient model for predicting potato tuber yield using RS and GIS techniques two vegetation indices, NDVI and soil adjusted vegetation index (SAVI), generated from images acquired by Landsat-8 and Sentinel-2 satellites were found to be highly effective in yield prediction [115]. In

addition, the indices enabled them to create maps of the study area that has clearly shown zones differing in productivity. This is very useful information for both farmers for implementing necessary management practices and for authorities in arriving at accurate production estimates. While manly researchers have used RS-based vegetation indices and GIS for predicting crop yields, several researchers have combined GIS with crop simulation or physiological models and demonstrated their strong performance in yield prediction [116–119]. Crop simulation models came into prominence due to their utility in designing management practices, assessing the role of climate variations on crop performance, and predicting yields [120, 121]. Similarly, physiological crop models have evolved from their original applications in farm management to measuring the impact of climatic changes on crop productivity. The ability to incorporate spatial variability of the inputs that go into physiological or simulation models makes them even more powerful for determining the interactions between climatic variation and crop productivity while highlighting the spatial heterogeneity. In a study that integrated RS data, crop growth model, and GIS, it was found that yield estimates from RS images were more precise compared to another approach where GIS climate layers and soil attributes were integrated into Oryza 2000 rice crop model highlighting the superiority of combining RS, GIS, and crop model for estimating crop yields [118]. To capture the spatial variability of input variables and their influence on yield estimates, [116] have linked RS and GIS with a growth model of soybean. The results demonstrated spatial variability in simulated yield estimates and the variability was primarily attributed to soil characteristics and rainfall. The availability of such spatial patterns from the simulated yield estimates is very helpful in productivity estimates in areas prone to abiotic stresses, for example, droughts, as well as providing insights into factors contributing to yield. Efforts also exist that have created web-based decision support systems based on a combination of simulation model and GIS towards making agronomic decisions [119]. In yet another approach, the Erosion Productivity Impact Calculator (EPIC), a model for the analysis of the relationship between soil erosion and crop yield at field level, was integrated with GIS and an Inference Engine (IE) towards global estimation of crop productivity [122]. While the integration of GIS expands the application of EPIC to regional or global level, the availability of IE helps in determining potential crop combinations for given growing conditions. This study not only demonstrated the ability of GIS-based EPIC for crop productivity simulations at global level but also delivered predictions for future yields and how they are adversely affected by global climate change underscoring the importance of the development of climate-resilient varieties of crops. Since traditional crop productivity simulations are based on site-specific crop models, [123] developed an operational crop model that can be utilized at the regional level, North China, by integrating USDA EPIC model with NASA MODIS LAI product from Earth Resources Observation System (EROS), ancillary ground data, and GIS [123]. Applications also exist where a combination of GIS and RS was used for assessing damage in some high-value crops. Cranberry is one such crop that exhibits extreme crop yield variations due to soil characteristics which in turn influence water and nutrient availability. Using GIS, GPS, and RS, [124] have created a spatial variation map for the crop enabling the analysis of crop losses within zones in a field or at the whole field level.

3.6 Precision farming

Precision Farming, also called Precision Agriculture (PA) or site-specific crop management (SSCM), is the application of technologies and principles to manage spatial and temporal variability associated with all aspects of agricultural production [125]. Earl et al. [126] defined it as a system that integrates information with crop production that is designed to increase long-term, site-specific as well as whole farm production efficiency, productivity, and profitability while minimizing unintended impacts on wildlife and the environment [126]. The operational goals of precision farming include better management of inputs such as seeds, fertilizers, pesticides, herbicides, and water using right amounts of inputs at the right place, and at the right time. Several crucial tools and systems such as GPS, GIS, and RS are required for the collection of timely geospatial information on soil-plant-animal requirements towards mining insights followed by leveraging those insights for prescribing and applying site-specific treatments towards improving agricultural productivity while contributing to sustainability and protecting the environment [127–129]. The role of different tools and technologies as well as the applications of precision farming are described in several review articles and references therein [9, 128, 130]. While GPS, GIS, and RS are vital for obtaining and analyzing the data for deriving insights, a key technology that implements the precision applications by leveraging the input of these three tools is variable rate technology (VRT). VRT systems take all the required information about a field such as soil maps, yield, infestation of pests, diseases, and weeds, and they determine the quantities of fertilizers, pesticides, herbicides, and other inputs and ensure their application at the right place and at the right time saving the input costs. The integration of GIS, GPS, and VRT technologies thus provides farmers an unprecedented ability to view field maps and apply input where and when needed towards ensuring crop productivity. Precision farming can be broadly divided into three steps or stages depending on data collection or site-directed or specific activities happening during, before, and after the crop growth period [131]. These are Preparatory or Pre-planting stage, Crop growth stage, and Harvesting stage. Role of GIS in each of these stages is discussed below.

Preparatory stage: This is essentially a planning stage that encompasses the data collection prior to planting and includes, among other things, gathering data on soil nutrient status, groundwater, previous crops and their residual influence on the next crops, and data on pests, diseases, and problematic weeds in that area that could affect yields of the crop to be planted. Data gathered on all these aspects are stored in a GIS system. Historical crop data from GIS also helps in 'variable planting' decisions that determine where to plant what crop/variety and to what extent so that the variable planting plan can be carried out in the field automatically by seeding machines [131]. A major activity in the preparatory stage is soil mapping and accurate prediction of soil properties is critical for precision farming interventions and for sustainable agriculture. The approach used for traditional soil mapping relies on a representative soil property from a location of focus and, therefore, has the limitation of not capturing the variability of soil properties in the maps generated. Moreover, soil sample collection, analysis, interpretation, and map generation are all monotonous, time-consuming, and costly. These methods have been improved greatly with the developments in spatial science and the geospatial modules in ArcGIS and other tools are being used extensively for their ability in spatial interpolation [132–134].

Crop growth stage: In this stage, the insights and the data gathered in the preparatory stage are retrieved through GIS and used for formulating and implementing management practices addressing irrigation, soil fertility, and protection from biotic and abiotic stresses. The availability of diverse types of imaging sensors, satellite and proximal sensing capabilities, GPS and GIS systems have enabled the assessment of plant characteristics, growth, health, and also in gathering information on the soils and infestation of pests and diseases, all with precise geospatial information. By using RGB, hyper- or multi- spectral remote sensing options for measuring the reflectance, fluorescence, or other useful emissions, it became possible to assess the disease status of the crops and take precise management measures for controlling the diseases at precise locations based on GPS and GIS information, thus improving crop health, and helping to improve productivity [135]. The use of advanced geoinformatics tools and data-driven recommendations are also being used for monitoring and managing pesticides and other plant protection inputs with a particular focus on sustainable practices towards improving productivity while reducing impact on the environment. Geoinformatics-based tools are also being leveraged for accelerating the crop germplasm that is suited for a given location or region towards realizing increased productivity as well as the germplasm that has tolerance to biotic or abiotic resistance needed for a given geographic location [136]. Spectral reflectance and vegetation indices of the crops being monitored combined with GIS are also playing a crucial role in managing nutrients and water stress in precision farming programs. Use of remote sensing and GIS in detecting nutrient stress of the crops enables site-specific correction management of nutrients thereby promoting plant growth while reducing the cost of cultivation due to targeted use of nutrients [137]. Similar site-specific management can also be implemented for addressing water stress and such an approach is especially attractive for croplands with limited water resources. The role of data analytics used in remote sensing, GIS, and other tools for achieving the goals of precision agriculture needs special mention. Early and accurate detection of biotic (pests and diseases) and abiotic (water deficiency) stresses is a prerequisite for taking appropriate management measures. Recent years have seen the use of several machine learning methods for developing models that are enabling both early and accurate detection of biotic stress agents such as diseases and weeds thereby helping precision crop protection [138].

Harvesting stage: This final stage serves the dual purpose of knowing the final output (yield) resulting from the season-long precision practices on the crop and leveraging this yield data to formulate a strategy for the next crop season [131]. The data collected at the harvesting stage is loaded into GIS for analysis towards generation of maps and insights for future use emphasizing the central role of GIS in precision farming. Yield monitoring and mapping is an important component of precision farming. The purpose of a yield monitor is to provide the farmer or a researcher with an accurate assessment of yield variability in the field and when combined with GPS, it can provide the data for creating yield maps [130]. Information on yield measurements and the geospatial context is critical for precision farming as it helps in formulating necessary tweaks to the management decisions for the next crop season.

3.7 Biomass assessment

Renewable sources of energy are crucial to achieving climate change and sustainability goals. Agricultural residues are a promising source of biomass-based energy the demand for which is rapidly increasing around the globe. One challenge with agricultural residues for efficiently channeling them for energy production is the fact that their availability is seasonal and is geographically widely distributed. A solution that can address this spatio-temporal variability, seasonal fluctuations in biomass supply levels, and identification and transport of residues to power plants is a critical prerequisite for biomass-based energy generation. GIS, in combination with remote sensing, can be a great tool for precise identification and assessment of the crop residues and for planning a given region's feedstock material for renewable energy and its economical transportation to power plants. GIS-based estimation of bioenergy potential enables a technologically advanced solution for leveraging the residues from existing cropping practices that promise even more benefits as the farmers shift from conventional to smart farming [139]. Some of the efforts in leveraging GIS and its partner technologies to this end are discussed below. Methods that can predict biomass potentials of a given region containing weather and crop production variations are of high value for enabling an efficient supply chain from biomass to power plants. By using BioSTAR, a carbon-based crop model, [140] have calculated biomass potentials for maize, triticale, and cup plant, and linked them with a GIS map of the soil dataset of Hannover region in Germany and demonstrated the utility of this method for predicting agricultural potentials under diverse environmental and crop management practices and conditions [140]. In a study that mapped rice cropland in a rural area in India, images from WorldView-2 satellite were used and the resulting map along with agricultural production statistics was analyzed in GIS for assessing the availability of rice straw as a feedstock for generating bioenergy [141]. In addition, the study also estimated the annual rice straw availability and the electrical power it could generate, thus providing valuable information for energy developers and policymakers for planning. Since the success and sustainability of a biomass-based energy generation project depend on several factors that include the feedstock resource, logistics, and environmental considerations, the role and value of GIS and key associated tools and technologies need to be understood prior to establishing the supply chain and the power plants. Two tools can help to address this task: GIS and life cycle assessment (LCA). While GIS is critical for assessing the resources dispersed in small or large areas, LCA is useful in evaluating the environmental impacts of bioenergy production projects. A comprehensive review on the application of LCA, especially spatial LCA, in understanding the impact of biomass-based energy generation on different ecosystem services and the value of integrating LCA and GIS to conduct a holistic assessment of environmental benefits in connection with bioenergy production recommended the inclusion of LCA as an essential component in planning bioenergy projects [142]. To assess the spatial and temporal availability of crop residues and to pinpoint locations for ideal power plants along with cost considerations, an integrated GIS-based biomass, site optimization, and logistics cost model was developed by using soil erosion, soil conditioning index (SCI), and crop residue yield indicators [143]. To estimate crop residues, prediction models based on artificial neural networks (ANNs) were developed for each of these indicators and were implemented on a GIS platform. The utility of this model was also demonstrated using a sustainable assessment of cotton stalks (CS) that are used to produce fuel pellets. An advantage of this model is that its use can be extended to assessment of multiple types of crop residues [143]. Models based on GIS and multi-criteria inclusion-exclusion analysis and facility location-allocation were also developed for the identification of sustainable crop biomass at larger spatial and longer temporal scales and to suggest ideal biogas plants along with cost considerations for biomass delivery [144].

3.8 Supply chain management

GIS technology has proved to be of great value in understanding and optimizing agricultural supply chains and its use is being extended to diverse crops and locations. For ease of understanding its impact on supply chains can be discussed using the following three categories.

3.8.1 Improving supply chain management process

GIS technology has the potential to assist the successful transition of traditional agriculture systems to smart systems. While there are many studies that have thoroughly investigated the role of big data analytics in supply chains of diverse industries, such studies are lacking in the application of big GIS analytics (BGA) in agriculture. To this end, a systematic review of recent literature examined the role of BGA in agricultural applications and has proposed a framework for supply chains where BGA can play even a bigger role in improving the quality of GIS applications in agriculture [145]. The proposed framework serves as a useful reference for scientists and authorities for the successful management of big GIS data and leveraging it for improving productivity. The utility of Geographical Information Technologies (GITs) for improving the complex supply chain management process in the cotton crop was explored and was found to be of great value since the GITs framework enables visualization of current states as well as alternative options and what-if analyses for all steps that require decision making [146]. Another important application for which GIS was used is the analysis of supply chain patterns and description of spatial components of safe crop product (SCP) in China [147]. By using the spatial functions provided by GIS such as representation, location, analysis, traceability coding, and other techniques, tracing and retracing of the quality of safe crop product (SCP) was achieved. This system was also successfully demonstrated in a real supply chain for (re)tracing of SCP. By developing a GIS-based constrained linear programming model for minimizing transportation and storage costs for soybean and its byproducts, and further optimizing this model using General Algebraic Modeling System (GAMS), [148] have identified the lowest cost supply chains. The origin to destination cost matrices and geographic data maps required for the model development and optimization was developed by ArcGIS Network Analyst and ArcMap, respectively. This study demonstrated the combinatorial utility of ArcGIS, ArcMap, and GAMS for developing optimal supply chains that are of value to the players in the process [148].

3.8.2 Decision support systems

Production of biofuels from renewable sources such as agricultural residues can reduce the usage of fossil fuels thereby helping in the reduction of greenhouse gases. Identification of ideal locations for establishing biofuel facilities and designing a cost-effective supply chain for transferring biomass to the facility is highly desirable. To this end, in one approach a decision support system (DSS) has been developed by integrating a GIS-based method and two modeling methods, simulation, and optimization [149]. While GIS-based method was used for selecting facility sites, the selected sites were run through simulation and optimization modeling, and together these three methods provided an integrated DSS for assessing the cost, energy use, and emissions for the facility candidates as well as minimizing supply chain costs. In another approach, an intelligent spatial decision support system (ISDSS) was proposed to overcome the drawbacks of GIS in enabling creation of a knowledge base that supports decision making. The ISDSS combines GIS and intelligent systems and has spatial data mining capability through IoT devices [150].

3.8.3 Locating power plants and developing supply chains

The sustainability of a biomass-based power plant depends on, among other things, a consistent supply of the feedstock, an economical supply chain, and an optimal location of the facility. The GIS-based analysis enables the identification of an ideal location for the plant and in making valid decisions related to the supply chain development. Using open-source GIS software, Latterini et al. [151], have simulated the identification of suitable locations for a small size power plant in Lazio region of Italy that can use olive prunings as the feedstock. This user-friendly and low-cost procedure, which can also be extended to other feedstocks, also provided supply chain costs for the evaluation of different sites and can serve as a useful tool for stakeholders in the development of economical biomass-based end-to-end supply chains [151]. In another study, an integrated approach combining GIS-based analysis with optimization modeling was developed resulting in a support system for decision-makers in comparing facility candidates and in minimizing supply chain costs [152]. The system developed could also be used for similar supply chains such as low capital biodiesel plants.

4. Conclusions

The use of GIS in agriculture has increased at a rapid pace during the recent decades and the number of applications and the prominence of GIS has further amplified in the recent years due to advances in digital technologies that have been leveraging GIS as an essential partner technology for assessing crops, soils, and their environments. As discussed in this chapter, GIS is being used at all stages of agricultural value chain. In addition to the historical, current, and popular uses of GIS in land suitability/use planning and management of water, soil, and biotic and abiotic stresses, the advent of digital agricultural tools and technologies has increasingly leveraged the capabilities of GIS in new and emerging applications in high fidelity crop monitoring, yield prediction, precision farming, and supply chain management for both primary produce and biomass utilization towards energy production. The multitude of capabilities and insights provided by GIS, including the recent enhancements to collect and analyze data in real time, has further elevated its importance in providing location/spatial intelligence needed for improving the productivity and profitability of farms through precision practices. With the current and emerging applications, in combination with existing and newer partner technologies, GIS has a pivotal role in achieving sustainable agricultural productivity.

Acknowledgements

We would like to thank Anu Swatantran for the support and helpful discussions, Chris Seifert, Andrea Arias, and Brian Lutz for support and funding, and Jochen Scheel, Steve Callistein, Janae Lehman Bell, and Nicole Janovick for rapid review and approval of the manuscript.

Conflict of interest

The authors declare no conflict of interest.

Author details

Parmita Ghosh[1*†] and Siva P. Kumpatla[2†]

1 Department of Digital Solutions Data Science, Corteva Agriscience™, Telangana, India

2 Department of Data Science and Bioinformatics, Corteva Agriscience™, Johnston, United States

*Address all correspondence to: parmita.ghosh@corteva.com

† These authors contributed equally.

References

[1] FAO. The Future of Food and Agriculture—Trends and Challenges. Rome: FAO; 2017

[2] Gomiero T. Soil degradation, land scarcity and food security: Reviewing a complex challenge. Sustainability. 2016; **8**(3):281

[3] Delgado JA, Short NM Jr, Roberts DP, Vandenberg B. Big data analysis for sustainable agriculture on a geospatial cloud framework. Frontiers in Sustainable Food Systems. 2019;**3**:54

[4] Burrough PA, McDonnell RA. Principles of Geographical Information Systems. Oxford: Oxford University Press; 1998

[5] Gomiero T. Alternative land management strategies and their impact on soil conservation. Agriculture. 2013; **3**(3):464-483

[6] Delgado JA, Vandenberg B, Neer D, D'Adamo R. Emerging nutrient management databases and networks of networks will have broad applicability in future machine learning and artificial intelligence applications in soil and water conservation. Journal of Soil and Water Conservation. 2019;**74**(6):113A-118A

[7] Burrough PA. Principles of Geographical: Information Systems for Land Resource Assessment. Oxford: Clarendon Press; 1986

[8] Soomro TR. GIS enabling smart agriculture. Smart Agriculture: An Approach towards Better Agriculture Management. 2015;**8**:1-6

[9] Gebeyehu MN. Remote sensing and GIS application in agriculture and natural resource management. International Journal of Environmental Sciences & Natural Resources. 2019; **19**(2):45-49

[10] Kingra PK, Majumder D, Singh SP. Application of remote sensing and GIS in agriculture and natural resource management under changing climatic conditions. Agricultural Research Journal. 2016;**53**(3):295-302

[11] Kazemi H, Akinci H. A land use suitability model for rainfed farming by Multi-criteria Decision-making Analysis (MCDA) and Geographic Information System (GIS). Ecological Engineering. 2018;**116**:1-6

[12] Kumar S, Karaliya SK, Chaudhary S. Precision farming technologies towards enhancing productivity and sustainability of rice-wheat cropping system. International Journal of Current Microbiology and Applied Sciences. 2017;**6**(3):142-151

[13] Sishodia RP, Ray RL, Singh SK. Applications of remote sensing in precision agriculture: A review. Remote Sensing. 2020;**12**(19):3136

[14] Weiss M, Jacob F, Duveiller G. Remote sensing for agricultural applications: A meta-review. Remote Sensing of Environment. 2020;**263**: 111402

[15] Chen Y, Yu J, Khan S. Spatial sensitivity analysis of multi-criteria weights in GIS-based land suitability evaluation. Environmental Modelling and Software. 2010;**25**(12):1582-1591

[16] Zolekar RB, Bhagat VS. Multi-criteria land suitability analysis for agriculture in hilly zone: Remote sensing and GIS approach. Computers and Electronics in Agriculture. 2015;**118**: 300-321

[17] Pan G, Pan J. Research in crop land suitability analysis based on GIS. In: International Conference on Computer and Computing Technologies in Agriculture. Berlin, Heidelberg: Springer; 2011. pp. 314-325

[18] El Baroudy AA. Mapping and evaluating land suitability using a GIS-based model. Catena. 2016;**140**:96-104

[19] Zabihi H, Ahmad A, Vogeler I, Said MN, Golmohammadi M, Golein B, et al. Land suitability procedure for sustainable citrus planning using the application of the analytical network process approach and GIS. Computers and Electronics in Agriculture. 2015;**117**: 114-126

[20] Tashayo B, Honarbakhsh A, Akbari M, Eftekhari M. Land suitability assessment for maize farming using a GIS-AHP method for a semi-arid region. Iran. Journal of the Saudi Society of Agricultural Sciences. 2020;**19**(5): 332-338

[21] Mendas A, Delali A. Integration of MultiCriteria Decision Analysis in GIS to develop land suitability for agriculture: Application to durum wheat cultivation in the region of Mleta in Algeria. Computers and Electronics in Agriculture. 2012;**83**:117-126

[22] Zhang J, Su Y, Wu J, Liang H. GIS based land suitability assessment for tobacco production using AHP and fuzzy set in Shandong province of China. Computers and Electronics in Agriculture. 2015;**114**:202-211

[23] Singha C, Swain KC. Land suitability evaluation criteria for agricultural crop selection: A review. Agricultural Reviews. 2016;**37**:2

[24] Yearbook FS. World Food and Agriculture. Rome: Food and Agriculture Organization of the United Nations; 2013. p. 15

[25] Unies N. World Population Prospects. The 2015 Revision, New York, United Nations. Population Division. 2015

[26] Consoli S, Vanella D. Mapping crop evapotranspiration by integrating vegetation indices into a soil water balance model. Agricultural Water Management. 2014;**143**:71-81

[27] Peragón JM, Delgado A, Pérez-Latorre FJ. A GIS-based quality assessment model for olive tree irrigation water in southern Spain. Agricultural Water Management. 2015; **148**:232-240

[28] Teeuw RM. Groundwater exploration using remote sensing and a low-cost geographical information system. Hydrogeology Journal. 1995; **3**(3):21-30

[29] Brahmabhatt VS, Dalwadi GB, Chhabra SB, Ray SS, Dadhwal VK. Land use/land cover change mapping in Mahi canal command area, Gujarat, using multi-temporal satellite data. Journal of the Indian Society of Remote Sensing. 2000;**28**(4):221-232

[30] Quan RS, Liu M, Lu M, Zhang LJ, Wang JJ, Xu SY. Waterlogging risk assessment based on land use/cover change: A case study in Pudong New Area, Shanghai. Environmental Earth Sciences. 2010;**61**(6):1113-1121

[31] Elhag M. Evaluation of different soil salinity mapping using remote sensing techniques in arid ecosystems, Saudi Arabia. Journal of Sensors. 2016;**1**:2016

[32] Engman ET, Gurney RJ. Remote Sensing in Hydrology. London: Chapman and Hall; 1991. p. 225

[33] Meijerink AMJ. Groundwater. In: Schultz GA, Engman ET, editors. RemoteSensing in Hydrology and Water Management. Berlin: Springer; 2000. pp. 305-325

[34] Pinder GF. Groundwater Modeling using Geographical Information Systems. John Wiley & Sons; 2002 Nov 6

[35] Chenini I, Mammou AB. Groundwater recharge study in arid region: An approach using GIS techniques and numerical modeling. Computational Geosciences. 2010;**36**(6): 801-817

[36] Tripathi MP, Panda RK, Raghuwanshi NS. Identification and prioritisation of critical sub-watersheds for soil conservation management using the SWAT model. Biosystems Engineering. 2003 Jul 1;**85**(3):365-379

[37] Singh, A.K., Prakash S.R., 2002. An integrated approach of remote sensing, geophysics and GIS to evaluation of groundwater potentiality of Ojhalasubwatershed, Mirzapur district, U.P., India

[38] Sikdar PK, Chakraborty S, Adhya E, Paul PK. Land use/land coverchanges and groundwater potential zoning in and around Raniganj coal miningarea, Bardhaman District, West Bengal: A GIS and remote sensing approach. Journal of Spatial Hydrology. 2004;**4**(2):1-24

[39] Singh AK, Raviprakash S, Mishra D, Singh S. Groundwater potential modeling in Chandraprabha subwatershed, UP using remote sensing, geoelectrical and GIS. Map India 2002, GIS Development. 2002:1-6

[40] Bhuyan SJ, Marzen LJ, Koelliker JK, Harrington JA Jr, Barnes PL. Assessment of runoff and sediment yield using remote sensing, GIS, andAGNPS. Journal of Soil and Water Conservation. 2001; **57**(6):35-41

[41] Aldababseh A, Temimi M, Maghelal P, Branch O, Wulfmeyer V. Multi-criteria evaluation of irrigated agriculture suitability to achieve food security in an arid environment. Sustainability. 2018;**10**(3):803

[42] Zolfaghary P, Zakerinia M, Kazemi H. A model for the use of urban treated wastewater in agriculture using multiple criteria decision making (MCDM) and geographic information system (GIS). Agricultural Water Management. 2021 Jan 1;**243**:106490

[43] Fortes PS, Platonov AE, Pereira LS. GISAREG—A GIS based irrigation scheduling simulation model to support improved water use. Agricultural Water Management. 2005;**77**(1–3):159-179

[44] Singh A. Managing the water resources problems of irrigated agriculture through geospatial techniques: An overview. Agricultural Water Management. 2016;**174**:2-10

[45] Parnes R. Soil Fertility. A Guide to Organic and Inorganic Soil Amendments. 2013

[46] Marschner H. Mineral Nutrition of Higher Plants. London-UK: Academic Press; 2008. p. 889

[47] Velayutham M, Bhattacharyya T. Soil resource management. In natural resource management for agricultural production in India'. Special publication during International Conference on Management of Natural Resources for Sustainable Agricultural Production in the 21st Century. 2000 Feb 14. pp. 14-18

[48] Wang G, Cao F. Integrated evaluation of soil fertility in Ginkgo (Ginkgo biloba L.) agroforestry systems

in Jiangsu, China. Agroforestry systems. 2011 Sep;**83**(1):89-100

[49] Viscarra Rossel RA, Webster R. Predicting soil properties from the Australian soil visible-near infrared spectroscopic database. European Journal of Soil Science. 2012;**63**: 848-860

[50] Shen W, Ni Y, Gao N, Bian B, Zheng S, Lin X, et al. Bacterial community composition is shaped by soil secondary salinization and acidification brought on by high nitrogen fertilization rates. Applied Soil Ecology. 2016;**108**:76-83

[51] Yang MH, Mouazen AM, Zhao XM, Guo X. Assessment of a soil fertility index using visible and near-infrared spectroscopy in the rice paddy region of southern China. European Journal of Soil Science. 2019

[52] Malczewski J. GIS and Multicriteria Decision Analysis. New York: John Wiley & Sons Inc.; 1999

[53] Shumilov OI, Kasatkina EA, Mielikainen K, Timonen M, Kanatjev AG. Palaeovolcanos, solar activity and pine tree-rings from the Kola Peninsula (northwestern Russia) over the last 560 years. International Journal of Environmental Research. 2011;**5**(4):855-864

[54] Salehi E, Zebardast L, Yavri AR. Detecting forest fragmentation with morphological image processing in golestan national park in northeast of Iran. International Journal of Environmental Research. 2012;**6**(2): 531-536

[55] Feng XY, Luo GP, Li CF, Dai L, Lu L. Dynamics of ecosystem service value caused by land use changes in ManasRiver of Xinjiang, China. International Journal of Environmental Research. 2012;**6**(2):499-508

[56] Asproth V, Holmberg SC, Håkansson A. Decision support for spatial planning and management of human settlements. In: Lasker GE, editor. Advances in Support Systems Research. Vol. 5. Ontario, Canada: Windsor; 1999. pp. 30-39

[57] Jiang H, Eastman JR. Application of fuzzy measures in multi-criteria evaluation in GIS. International Journal of Geographical Information System. 2000;**14**:173-184

[58] Malczewski J, Chapman T, Flegel C, Walters D, Shrubsole D, Healy MA. GIS-multicriteria evaluation with ordered weighted averaging (OWA): Case study of developing watershed management strategies. Environment & Planning A. 2003;**35**(10):1769-1784

[59] Malczewski J, Rinner C. Exploring multicriteria decision strategies in GIS with linguistic quantifiers: A case study of residential quality evaluation. Journal of Geography System. 2005;**7**(2):249-268

[60] Kusre BC, Ghosh P, Nath K. Prioritization of soil conservation measures using erodibility indices as criteria in Sikkim (India). Journal of Earth System Science. 2018;**127**(6):1-3

[61] AbdelRahman MA, Natarajan A, Srinivasamurthy CA, Hegde R. Estimating soil fertility status in physically degraded land using GIS and remote sensing techniques in Chamarajanagar district, Karnataka, India. The Egyptian Journal of Remote Sensing and Space Science. 2016;**19**(1): 95-108

[62] Mokarram M, Hojati M. Using ordered weight averaging (OWA) aggregation for multi-criteria soil

fertility evaluation by GIS (case study: Southeast Iran). Computers and Electronics in Agriculture. 2017;**132**:1-3

[63] Li M, Zhang XL. GIS-based evaluation of farmland soil fertility and its relationships with soil profile configuration pattern. Ying Yong Sheng tai xue bao= The. Journal of Applied Ecology. 2011;**22**(1):129-136

[64] Leena HU, Premasudha BG, Basavaraja PK. Sensible approach for soil fertility management using GIS cloud. In: 2016 International Conference on Advances in Computing, Communications and Informatics (ICACCI). IEEE; 2016 Sep 21. pp. 2776-2781

[65] Mishra A, Pattnaik T, Das D, Das M. Soil fertility maps preparation using GPS and GIS in Dhenkanal District, Odisha, India. International Journal of Plant Soil Science. 2014;**3**(8):986-994

[66] Tunçay T, Kılıç Ş, Dedeoğlu M, Dengiz O, Başkan O, Bayramin İ. Assessing soil fertility index based on remote sensing and gis techniques with field validation in a semiarid agricultural ecosystem. Journal of Arid Environments. 2021;**190**:104525

[67] Liu HH, Zhao YM, Wang XY, Feng YJ, Yang W. Discussion of evaluation methods on soil fertility. Journal of Yangtze River Scientific Research Institute. 2008;**25**(3):62-66

[68] Nie Y, Yu J, Wu Y, Yu L, Jing Y, Zhou Y. A comprehensive evaluation of soil fertility of cultivated land: A GIS-based soil basic Niche- Fitness Model. Communications in Soil Science and Plant Analysis. 2016;**47**(5):670-678

[69] Ren Y, Chen J, Chen L, Zhang H, Zhang K. Research and implementation of a universal workflow model to evaluate the soil fertility based on OGC Web Service. Geo Spatial Information Science. 2018;**21**(4):346-357

[70] Williams H, Colombi T, Keller T. The influence of soil management on soil health: An on-farm study in southern Sweden. Geoderma. 2020;**360**:114010

[71] Li ZW, Huang JQ, Li YY, Guo W, Zhu JF. Assessment on soil fertility of Dongting Lake wetland area (China) based on GIS and fuzzy evaluation. Journal of Central South University. 2011;**18**(5):1465-1472

[72] Li C, Chen G, Zeng G, Ye J. The study of soil fertility spatial variation feature based on GIS and data mining. In: International Conference on Computer and Computing Technologies in Agriculture. Berlin, Heidelberg: Springer; 2012. pp. 211-220

[73] Lal R. Restoring soil quality to mitigate soil degradation. Sustainability. 2015;**7**:5875-5895

[74] Hunter MC, Smith RG, Schipanski ME, Atwood LW, Mortensen DA. Agriculture in 2050: Recalibrating targets for Sustainable intensification. Bioscience. 2017;**67**: 386-391

[75] El-Naggar A, Lee SS, Rinklebe J, Farooq M, Song H, Sarmah AK, et al. Biochar application to low fertility soils: A review of current status, and future prospects. Geoderma. 2019;**337**:536-554

[76] Schipanski M, Snapp S, Jackson LE. Ecologically based nutrient management. In: Snapp SS, Pound CB, editors. Agricultural Systems. second ed. Academic Press; 2017. pp. 203-257

[77] Bagherzadeh A, Gholizadeh A. Assessment of soil fertility for sugar beet production using fuzzy AHP approach

and GIS in the Northeastern region of Iran. Agricultural Research (Kurashiki). 2018;**7**(1):61-71

[78] Govers G, Merckx R, Van Wesemael B, Van Oost K. Soil conservation in the 21st Century: Why we need smart agricultural intensification. The Soil. 2017;**3**:45-59

[79] Kome GK, Enang RK, Palmer B, Yerima BPK. Knowledge and management of soil fertility by farmers in Western Cameroon. Geoderma Regional. 2018;**13**:43-51

[80] Yageta Y, Osbahr H, Morimoto Y, Clark J. Comparing farmers' qualitative evaluation of soil fertility with quantitative sol fertility indicators I Kitui Count, Kenya. Geoderma. 2019;**344**: 153-163

[81] Patil PL, Kuligod VB, Gundlur SS, Katti J, Nagaral IN, Shikrashetti P, et al. Soil fertility mapping in Dindur sub-watershed of Karnataka for site specific recommendations. Journal of the Indian Society of Soil Science. 2016;**64**(4):381-390

[82] Jackai LE, Daoust RA. Insect pests of cowpeas. Annual Review of Entomology. 1986;**31**(1):95-119

[83] Haile FJ, Higley LG, Specht JE. Soybean cultivars and insect defoliation: Yield loss and economic injury levels. Agronomy Journal. 1998;**90**(3): 344-352

[84] Cerda R, Avelino J, Gary C, Tixier P, Lechevallier E, Allinne C. Primary and secondary yield losses caused by pests and diseases: Assessment and modeling in coffee. PloS One. 2017 Jan 3;**12**(1): e0169133

[85] Ranjan R, Vinayak S. Application of remote sensing and GIS in plant disease management. TTPP. 2020;**509**

[86] Duarte F, Calvo MV, Borges A, Scatoni IB. Geostatistics applied to the study of the spatial distribution of insects and its use in integrated pest management. Revista Agronómica del Noroeste Argentino. 2015;**35**(2): 9-20

[87] Liu Y, Shi J. Predicting the potential global geographical distribution of two Icerya species under climate change. Forests. 2020;**11**(6):684

[88] Bale J, Walters KF. Overwintering biology as a guide to the establishment potential of non-native arthropods in the UK. In: Atkinson DA, Thorndyke M. Temperature and Development. 2001 Jan 1

[89] Pretty J, Benton TG, Bharucha ZP, Dicks LV, Flora CB, Godfray HC, et al. Global assessment of agricultural system redesign for sustainable intensification. Nature Sustainability. 2018;**1**(8):441-446

[90] Zeng J, Liu Y, Zhang H, Liu J, Jiang Y, Wyckhuys KA, et al. Global warming modifies long-distance migration of an agricultural insect pest. Journal of Pesticide Science. 2020;**93**(2): 569-581

[91] Carrière Y, Ellsworth PC, Dutilleul P, Ellers-Kirk C, Barkley V, Antilla L. A GIS-based approach for areawide pest management: The scales of Lygus hesperus movements to cotton from alfalfa, weeds, and cotton. Entomologia Experimentalis et Applicata. 2006;**118**(3):203-210

[92] Latchininsky AV, Sivanpillai R. Locust habitat monitoring and risk assessment using remote sensing and GIS technologies. In: Integrated Management of Arthropod Pests and Insect Borne Diseases. Dordrecht: Springer; 2010. pp. 163-188

[93] Mahlein AK. Plant disease detection by imaging sensors–parallels and specific demands for precision agriculture and plant phenotyping. Plant Disease. 2016; **100**(2):241-251

[94] Franke J, Menz G. Multi-temporal wheat disease detection by multi-spectral remote sensing. Precision Agriculture. 2007;**8**(3):161-172

[95] Mahlein AK, Rumpf T, Welke P, Dehne HW, Plümer L, Steiner U, et al. Development of spectral indices for detecting and identifying plant diseases. Remote Sensing of Environment. 2013; **128**:21-30

[96] Zhang J, Huang Y, Yuan L, Yang G, Chen L, Zhao C. Using satellite multispectral imagery for damage mapping of armyworm (Spodoptera frugiperda) in maize at a regional scale. Pest Management Science. 2016;**72**(2): 335-348

[97] Chau VN, Holland J, Cassells S, Tuohy M. Using GIS to map impacts upon agriculture from extreme floods in Vietnam. Applied Geography. 2013;**41**: 65-74

[98] Consuegra D, Joerin F, Vitalini F. Flood delineation and impact assessment in agricultural land using GIS technology. In: Geographical Information Systems in Assessing Natural Hazards. Dordrecht: Springer; 1995. pp. 177-198

[99] Yamada Y. Detection of flood-inundated area and relation between the area and micro-geomorphology using SAR and GIS. In: IGARSS 2001. Scanning the Present and Resolving the Future. Proceedings. IEEE 2001 International Geoscience and Remote Sensing Symposium (Cat. No. 01CH37217) IEEE. 2001 Jul 9;**7**:3282-3284

[100] Pistrika A. Flood damage estimation based on flood simulation scenarios and a GIS platform. European Water. 2010;**30**:3-11

[101] Murad H, Islam AK. Drought assessment using remote sensing and GIS in north-west region of Bangladesh. In: Proceedings of the 3rd international conference on water & flood management. 2011 Jan 8. pp. 797-804

[102] Suryabhagavan KV. GIS-based climate variability and drought characterization in Ethiopia over three decades. Weather and Climate Extremes. 2017;**15**:11-23

[103] Saha A, Patil M, Goyal VC, Rathore DS. Assessment and impact of soil moisture index in agricultural drought estimation using remote sensing and GIS techniques. In Multidisciplinary digital publishing institute proceedings 2018;**7** (1):2

[104] Reynolds CA, Yitayew M, Slack DC, Hutchinson CF, Huete A, Petersen MS. Estimating crop yields and production by integrating the FAO Crop Specific Water Balance model with real-time satellite data and ground-based ancillary data. International Journal of Remote Sensing. 2000;**21**(18):3487-3508

[105] Dadhwal VK, Ray SS. Crop assessment using remote sensing-Part II: Crop condition and yield assessment. Indian Journal of Agricultural Economics. 2000;**55**:55-67

[106] Taylor JC, Wood GA, Thomas G. Mapping yield potential with remote sensing. Precision Agriculture. 1997;**1**: 713-720

[107] Xue J, Su B. Significant remote sensing vegetation indices: A review of developments and applications. Journal of sensors. 2017;**23**

[108] Adhav V, Das S, Patkar V. Crop health monitoring using geospatial technologies for nashik district, Maharashtra. International Journal of Modern Agriculture. 2021;**10**(2): 1395-1409

[109] Lowder SK, Sánchez MV, Bertini R. Which farms feed the world and has farmland become more concentrated? World Development. 2021;**142**:105455

[110] Brewer K, Clulow A, Sibanda M, Gokool S, Naiken V, Mabhaudhi T. Predicting the chlorophyll content of maize over phenotyping as a proxy for crop health in smallholder farming systems. Remote Sensing. 2022;**14**(3):518

[111] Vanlauwe B, Coyne D, Gockowski J, Hauser S, Huising J, Masso C, et al. Sustainable intensification and the African smallholder farmer. Current Opinion in Environment Sustainability. 2014;**8**:15-22

[112] Singhal G, Bansod B, Mathew L. Real time Crop health monitoring using Remote Sensing and ancillary information using GIS. In: 19th Esri India User Conference. India; 2018

[113] Maloom JM, Saludes RB, Dorado MA, Cruz PC. Development of a GIS-based model for predicting rice yield. Philiphines Journal of Crop Science. 2014;**39**:8-19

[114] Zinna AW, Suryabhagavan KV. Remote sensing and GIS based spectro-agrometeorological maize yield forecast model for South Tigray Zone, Ethiopia. Journal of Geographic Information System. 2016;**8**(2):282-292

[115] Al-Gaadi KA, Hassaballa AA, Tola E, Kayad AG, Madugundu R, Alblewi B, et al. Prediction of potato crop yield using precision agriculture techniques. PLoS One. 2016;**11**(9): e0162219

[116] Carbone GJ, Narurnalani S, King M. Application of remote sensing and technologies with physiological crop models. Photogrammetric Engineering and Remote Sensing. 1996;**62**(2):171-179

[117] Hartkamp AD, White JW, Hoogenboom G. Interfacing geographic information systems with agronomic modeling: A review. Agronomy Journal. 1999;**91**:761-772

[118] Sailaja B, Voleti SR, Subrahmanyam D, Nathawat MS, Rao NH. Regional rice yield estimation by integration of spatial technologies and crop model. Journal of Remote Sensing & GIS. 2013;**4**(2):56-66

[119] Kadiyala MD, Nedumaran S, Singh P, Chukka S, Irshad MA, Bantilan MC. An integrated crop model and GIS decision support system for assisting agronomic decision making under climate change. The Science of the Total Environment. 2015;**521**:123-134

[120] Bouman BAM, Van Keulen H, Van Laar HH, et al. The 'School of de Wit' Crop Growth Simulation Models: Pedigree and Historical Overview. Agricultural Systems. 1996;**52**:171-198

[121] Jones JW, Hoogenboom G, Porter CH, et al. The DSSAT Cropping System Model. European Journal of Agronomy. 2003;**18**:235-265

[122] Tan G, Shibasaki R. Global estimation of crop productivity and the impacts of global warming by GIS and EPIC integration. Ecological Modelling. 2003;**168**(3):357-370

[123] Yang P, Tan GX, Zha Y, Shibasaki R. Integrating remotely sensed data with an ecosystem model to estimate crop

yield in north China. In: Proceedings of XXth ISPRS Congress Proceedings Commission VII, WG VII/2, Istanbul, Turkey; 2004 Jul. pp. 150-156

[124] Oudemans PV, Pozdnyakova L, Hughes MG, Rahman F. GIS and remote sensing for detecting yield loss in cranberry culture. Journal of Nematology. 2002;**34**(3):207

[125] Pierce FJ, Nowak P. Aspects of precision agriculture. Advances in Agronomy. 1999 Jan 1;**67**:1-85

[126] Earl R, Wheeler PN, Blackmore BS, Godwin RJ. Precision farming: The management of variability. The Journal of the Institution of Agricultural Engineers. 1996;**51**:18-23

[127] Strickland RM, Ess DR, Parsons SD. Precision farming and precision pest management: The power of new crop production technologies. Journal of Nematology. 1998;**30**(4):431

[128] Goswami SB, Matin S, Saxena A, Bairagi GD. A review: The application of Remote Sensing, GIS and GPS in Precision Agriculture. International Journal of Advanced Technology & Engineering Research. 2012;**2**(1):50-54

[129] Balafoutis A, Beck B, Fountas S, Vangeyte J, Wal TV, Soto I, et al. Precision agriculture technologies positively contributing to GHG emissions mitigation, farm productivity and economics. Sustainability. 2017;**9**(8):1339

[130] Banu S. Precision agriculture: Tomorrow's technology for today's farmer. Journal of Food Processing & Technology. 2015;**6**(8):1

[131] Pfister B. What is Precision Agriculture?. 1998. Available from: http://www.directionsmag.com/articles/what-is-precision-agriculture/124210

[132] Fasina AS, Omolayo FO, Ajayi OS, Falodun AA. Influence of land use on soil properties of three mapping units in Southwestern Nigeria-Implications for sustainable soil management. Research Journal of Applied Sciences. 2007;**2**(8): 879-883

[133] Pravat KS, Gouri B, Ramkrishna M. Spatial analysis of soil properties using GIS based geostatistics models. Earth System and Environment. 2016;**2**:107. DOI: 10.1007/s40808-016-0160-4

[134] Kingsley J, Lawani SO, Esther AO, Ndiye KM, Sunday OJ, Penížek V. Predictive mapping of soil properties for precision agriculture using geographic information system (GIS) based geostatistics models. Modern Applied Science. 2019;**13**(10)

[135] Oerke EC. Remote sensing of diseases. Annual Review of Phytopathology. 2020;**58**:225-252

[136] Roberts DP, Short NM, Sill J, Lakshman DK, Hu X, Buser M. Precision agriculture and geospatial techniques for sustainable disease control. Indian Phytopathology. 2021;**74**(2):287-305

[137] Shanmugapriya P, Rathika S, Ramesh T, Janaki P. Applications of remote sensing in agriculture-A Review. International Journal of Current Microbiology and Applied Sciences. 2019;**8**(1):2270-2283

[138] Behmann J, Mahlein AK, Rumpf T, Römer C, Plümer L. A review of advanced machine learning methods for the detection of biotic stress in precision crop protection. Precision Agriculture. 2015;**16**(3):239-260

[139] Bharti A, Paritosh K, Mandla VR, Chawade A, Vivekanand V. GIS application for the estimation of bioenergy potential from agriculture

residues: An overview. Energies. 2021; **14**(4):898

[140] Bauböck R, Karpenstein-Machan M, Kappas M. Computing the biomass potentials for maize and two alternative energy crops, triticale and cup plant (*Silphium perfoliatum* L.), with the crop model BioSTAR in the region of Hannover (Germany). Environmental Sciences Europe. 2014;**26**(1):1-2

[141] Hiloidhari M, Baruah DC. GIS mapping of rice straw residue for bioenergy purpose in a rural area of Assam, India. Biomass and Bioenergy. 2014;**71**:125-133

[142] Hiloidhari M, Baruah DC, Singh A, Kataki S, Medhi K, Kumari S, et al. Emerging role of Geographical Information System (GIS), Life Cycle Assessment (LCA) and spatial LCA (GIS-LCA) in sustainable bioenergy planning. Bioresource Technology. 2017; **242**:218-226

[143] Sahoo K, Hawkins GL, Yao XA, Samples K, Mani S. GIS-based biomass assessment and supply logistics system for a sustainable biorefinery: A case study with cotton stalks in the Southeastern US. Applied Energy. 2016; **182**:260-273

[144] Sahoo K, Mani S, Das L, Bettinger P. GIS-based assessment of sustainable crop residues for optimal siting of biogas plants. Biomass and Bioenergy. 2018;**110**:63-74

[145] Sharma R, Kamble SS, Gunasekaran A. Big GIS analytics framework for agriculture supply chains: A literature review identifying the current trends and future perspectives. Computers and Electronics in Agriculture. 2018;**155**:103-120

[146] Perveen S, Arsalan MH. A review of potential use of geo-information technologies for cotton supply chain management. International Journal of Biodiversity and Conservation. 2014; **6**(10):724-734

[147] Qu XH, Zhuang DF, Qiu DS. Studies on GIS based tracing and traceability of safe crop product in China. Agricultural Sciences in China. 2007;**6**(6):724-731

[148] Nardi MG, Sperry SE, Davis TD. Grain supply chain management optimization using ARCGIS in Argentina. In: 2007 ESRI User Conference Proceedings. Redlands, CA; 2007. pp. 1-23

[149] Zhang F, Johnson D, Johnson M, Watkins D, Froese R, Wang J. Decision support system integrating GIS with simulation and optimisation for a biofuel supply chain. Renewable Energy. 2016; **85**:740-748

[150] Yusianto R, Hardjomidjojo H. Intelligent spatial decision support system concept in the potato agro-industry supply chain. In: 2020 International Conference on Computer Science and Its Application in Agriculture (ICOSICA). Bogor, Indonesia: IEEE; 2020. pp. 1-7

[151] Latterini F, Stefanoni W, Suardi A, Alfano V, Bergonzoli S, Palmieri N, et al. A GIS approach to locate a small size biomass plant powered by olive pruning and to estimate supply chain costs. Energies. 2020;**13**(13):3385

[152] Zhang F, Wang J, Liu S, Zhang S, Sutherland JW. Integrating GIS with optimization method for a biofuel feedstock supply chain. Biomass and Bioenergy. 2017;**98**:194-205

Chapter 6

Multi-Criteria Land Suitability Analysis for Agriculture Using AHP and Remote Sensing Data of Northern Region India

Mujahid Ali Khan, Rizwan Ahmad and Haris Hasan Khan

Abstract

The purpose of this study was to identify adequate agricultural sites in Punjab's Northern region India district (India). This research employed the "Analytic Hierarchy Process (AHP)" approach, which is extensively used in land use appropriateness studies. Great soil type, land use, land cover, soil moisture, slope, aspect, elevation, drainage, geology, and geomorphology were all incorporated into the application. The ranks of influencing criteria were calculated using expert judgments and correlation analysis, while the weights were determined using a pairwise comparison matrix. The scores for sub-parameters with internal variations in the criteria assigned based on field work and published norms. The study area is considered to be highly appropriate for agricultural production in 41.2% (39044.28 ha), moderately suitable in 14.3% (13498.76 ha), and marginally suitable in 4.2% (3993 ha). Furthermore, it was discovered that 1.9% of the land is now unfit for agricultural production (1766.6 ha), while 38.4% of the area is permanently unsuitable (36372.6 ha). The following facts were also discovered to be important in achieving these results: a large portion (approximately 45%) of the study area is covered with forests, built-up areas, and water bodies, the soil depth is insufficient for agricultural production, the slope in the study area is quite steep, and thus the erosion degree is high.

Keywords: land suitability, AHP, FAO, northern region India, multi-criteria decision analysis, pairwise comparison matrix

1. Introduction

Agriculture, being man's most fundamental profession, has benefitted immensely from technological advancements ranging from shifting cultivation to high precision farming. With the advent of civilization, man learned about additional crops and began to produce a variety of crops. As the human population grew and civilization progressed, people began to dwell in one location and farm the same land year after year. Now that agriculture has evolved into a profession, it is known as commercial agriculture, with precision agriculture and sustainable agriculture as key components.

The world's population is rapidly increasing these days. To meet the rising demand for food, the agricultural community must produce more and more. Because it is difficult to bring additional land under cultivation (extensive farming) in the current scenario, when land is a limited issue, the agricultural community should take on the task of producing more and more food with the land that is available (intensive farming). On the contrary, growing worldwide concern for human health and the environment opposes the use of higher levels of pesticides and fertilizers, as well as genetically modified plants. The latter, on the other hand, are present technologies that have the potential to improve food production.

Crop needs and soil/land conditions influence adaptability. Suitability is determined by matching the land features to the crop needs. Suitability is a measure of whether a land unit's features meet the needs of a certain type of land use (FAO). Aside from land and soil qualities, additional driving elements that might impact crop choices include socioeconomic, market, and infrastructural factors.

The FAO Land Evaluation Framework is based on previous land capabilities methodologies. In this case, the overall land appropriateness of a land area for a particular land use is assessed using a series of more or less independent land attributes, each of which may limit the land-use potential. These assessments are frequently used to classify map units in natural resource inventories. A soil survey's legend categories are divided into suitability subclasses based on the quantity and severity of land use restrictions.

In the FAO framework, there are two sorts of categories based on the scale of measurement of appropriateness.

- Qualitative: in reconnaissance investigations, the classes are rated based on the physical production potential of the land. It is employed to assess environmental, social, and economic factors.

- Quantitative: the classes are specified in numerical terms that allow for comparison of the objectives. There are a lot of economic parameters employed here.

By introducing quantification of land suitability indicators over a whole area, quantified land evaluation [1] revolutionized land suitability evaluation. The area is divided into small grid cells, and cell-based modeling has started. The indicators, on the other hand, must be quantitative. Geographical information systems and geostatistical approaches are commonly used in such land suitability analyses.

The FAO Framework identifies four categories of increasing details, as shown in **Table 1**.

Land appropriateness is a factor in determining a land use's long-term viability. The sustainability of a land use is defined by its suitability and vulnerability. Maximum appropriateness and minimum vulnerability should be the goals of sustainable land use (**Figure 1**) [2].

According to Rossiter [3], land is distinctive in every location, and this uniqueness has an impact on land usage. He also mentions how land evaluation might help with agricultural support services.

The multi-criteria land suitability was evaluated in a non-spatial manner, assuming spatial homogeneity across the study region. However, in circumstances like land

S. No.	Categories	Explanation
1	Land Suitability Orders	Reflecting kinds of suitability
2	Land Suitability Classes	Reflecting degrees of suitability within Orders
3	Land Suitability Subclasses	Reflecting kinds of limitation, or main kinds of improvement measures required, within Classes
4	Land Suitability Units	Reflecting minor differences in required management

Table 1.
FAO structure of land suitability classification.

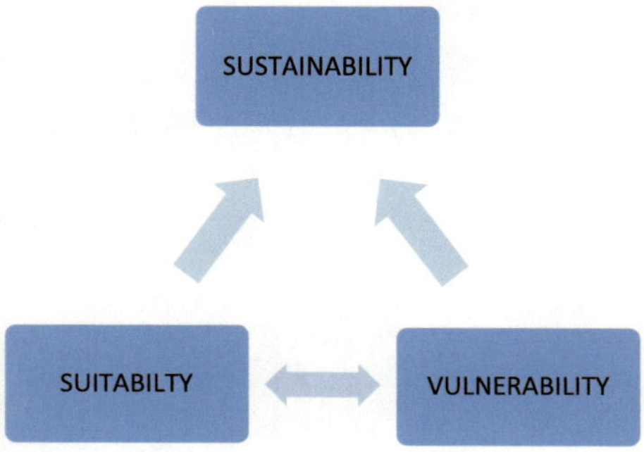

Figure 1.
Land use sustainability (after [2]).

suitability studies, when decisions are made based on factors that change over space, this is impossible [4]. Non-spatial traditional MCDM techniques average or total the effects that are judged appropriate for the entire area under consideration to address the spatial decision [5]. Jankowski [6] suggests that making, MCE, and GIS can all be combined. For many crops, MCE appears to be applicable in GIS-based land suitability analyses [7].

Ranking and rating are two widely used MCE approaches in land suitability evaluations. These methods lack a theoretical underpinning in determining the weights. The weights are assigned quite haphazardly in these procedures. They do not take into account comparisons between criteria and classifications. Furthermore, the results of such an investigation are grouped together using a simple Boolean overlay or weighted aggregation.

Since its beginnings, several researchers have examined the Analytic Hierarchy Process (AHP) [8]. The Analytic Hierarchy Process (AHP) is a method for making multi-criteria decisions (MCDM). The earliest reference we have identified is from 1972 [9]. The method was then discussed in detail in a paper published in the Journal of Mathematical Psychology [10]. The vast majority of applications continue to use AHP in the manner specified in this first article, oblivious to subsequent developments. This study draws out the significant trends in methodological advancements and future research in this vital topic.

AHP has been widely used since its introduction, for example, in flexible manufacturing system [11], Machine selection [12], industrial R&D project selection and resource allocation [13], Delphi method [14], Computer-aided machine-tool selection [15], evaluating machine tool alternatives [16], Integrating fuzzy theory and hierarchy concepts to evaluate software quality [17], product design in concurrent engineering [18]. Issue resolution for conceptual design using AHP [19]. Selection of appropriate schedule delay analysis method [20].

2. Study area and data used

2.1 Study area

Northern region India is a city in India's Punjab state. On July 27, 2011, Northern region India was formally designated as a district of Punjab state (Previously it was a Tehsil of Gurdaspur district, Punjab). Northern region India district is located in Punjab's northernmost region (**Figure 2**).

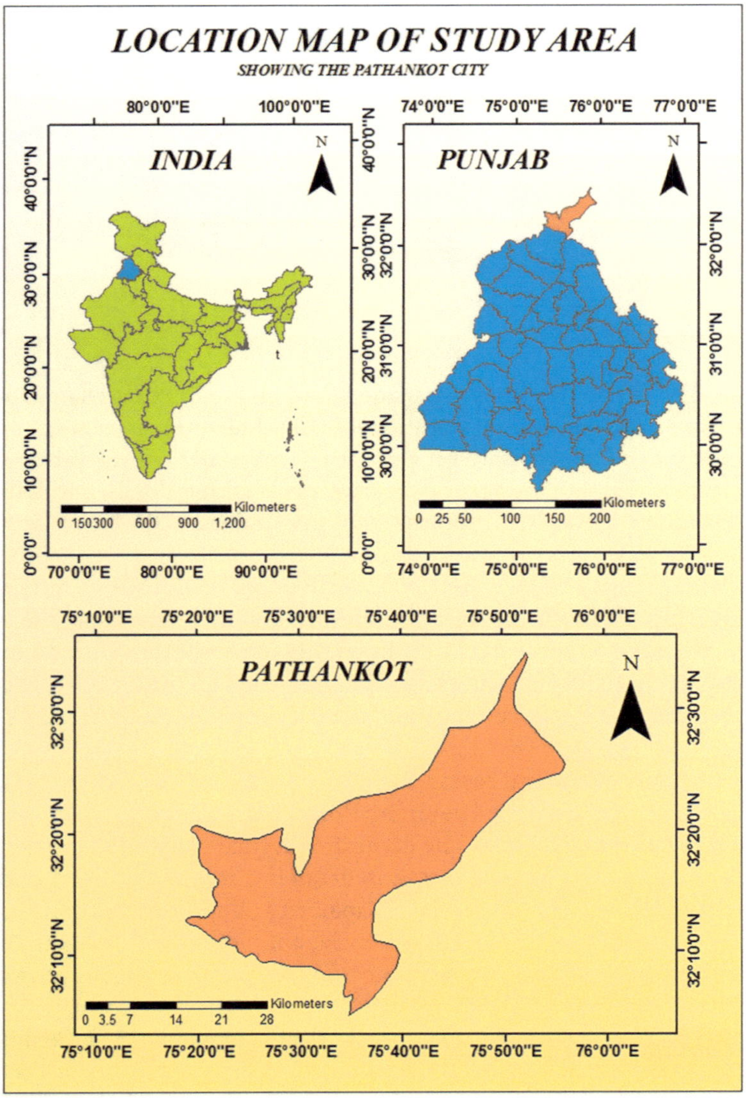

Figure 2.
Location of the study area.

It is where the three northern states of Punjab, Himachal Pradesh, and Jammu and Kashmir come together. Northern region India serves as a transportation hub for the three northern states due to its strategic location. It is the last city in Punjab on the national highway that connects Jammu and Kashmir to the rest of the country. Northern region India is also a major educational center for the nearby states of Jammu and Kashmir and Himachal Pradesh. It is located in the Jalandhar division, between the Ravi and the Beas rivers.

Northern region India district is located between 32°23′31″ and 32°23′52″ north latitudes and 75°39′55″ to 75°56′12″ east longitudes and covers an area of 27,123 ha. On a 1:50 K scale, the Survey of India 43 P/11, 43 P/14, and 43 P/15 top sheets cover the area.

2.2 Data used

See **Table 2**.

2.3 Data sets prepared

The shuttle radar topography mission (SRTM) elevation data (30 m resolution) obtained from USGS explorer (https://earthexplorer.usgs.gov/) was used to create a digital elevation model (DEM) of the study area. Using the DEM data slope, aspect, drainage density, Elevation thematic layers were built using ArcGIS 10.5. The land use and land cover data is downloaded by Nasa earth data ORNL DAAC (https://daac.ornl.gov/cgi-bin/dsviewer.pl?ds_id=1336). Soil map was obtained from European soil data center (ESDAC) which was published by National atlas and thematic map organization, Department of science and technology (https://esdac.jrc.ec.europa.eu/content/national-atlas-india-northern-india-plate-199-soil-regions).Geology and Geomorphology data of Northern region India is downloaded from Bhukosh (http://bhukosh.gsi.gov.in/Bhukosh/MapViewer.aspx). Sentinel2 data of the study area is downloaded from Copernicus(https://scihub.copernicus.eu/). Using the Sentinel 2 data NDWI was built in ArcGIS 10.5.These resulted thematic maps: Slope, LULC, NDWI and Drainage density were integrated in ArcGIS 10.5 and finally soil suitability map was obtained (**Figures 3–11; Tables 3–11**).

S. No	Data set	Spatial resolution/scale	Source
1	Sentinel 2	10 m	Copernicus
2	SRTM Dem	30 m	USGS Earth Explorer
3	Land use and land cover data	100 m	NASA Earth Data ORNL DAAC
4	Soil map	1:2000000	European soil data center (ESDAC)
5	Geology	1:2000000	Bhukosh
6	Geomorphology	1:250000	Bhukosh

Table 2.
Data set and data source.

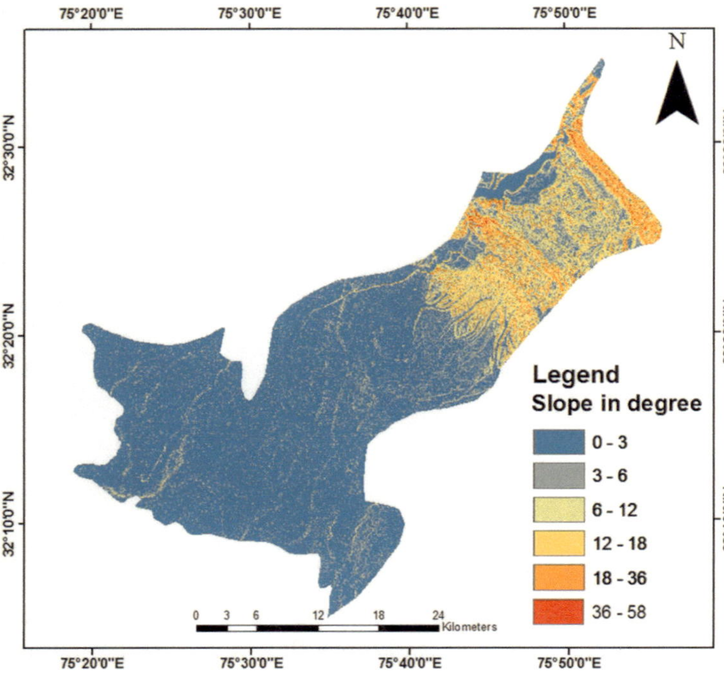

Figure 3.
Slope map.

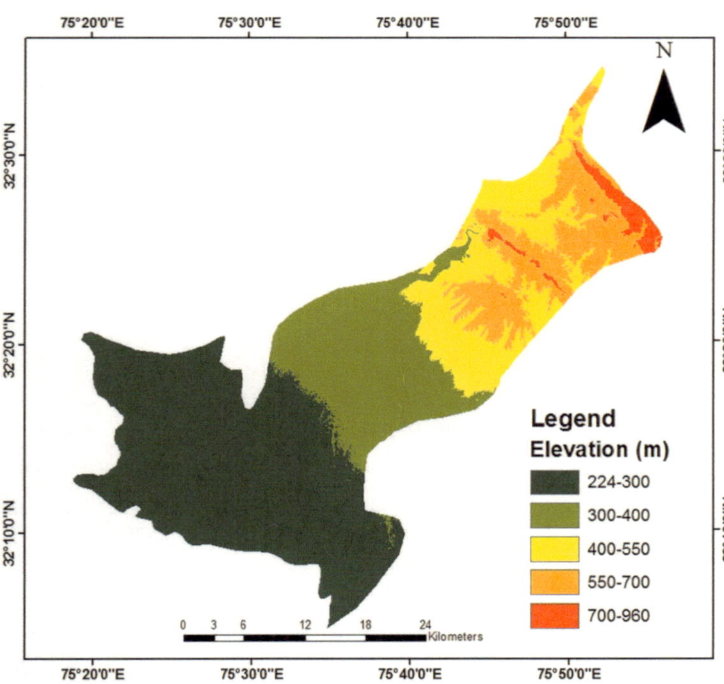

Figure 4.
Elevation map.

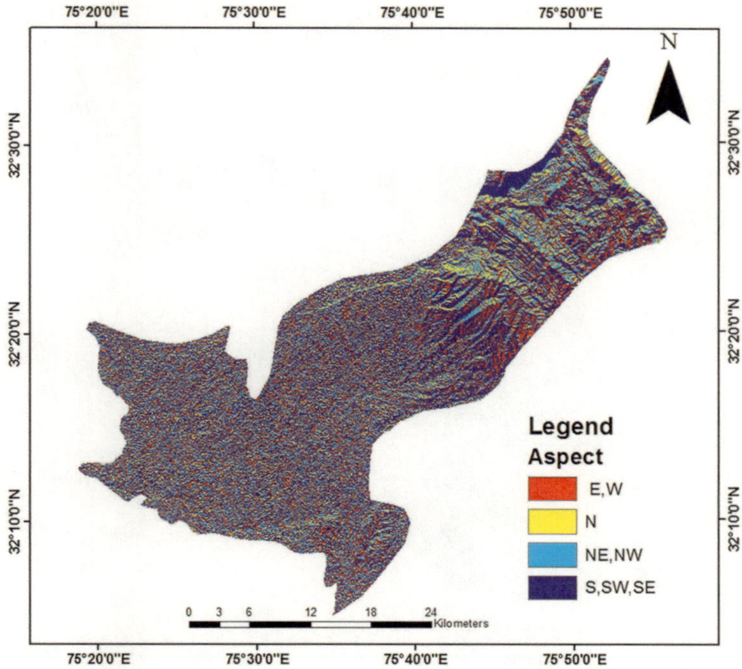

Figure 5.
Aspect map.

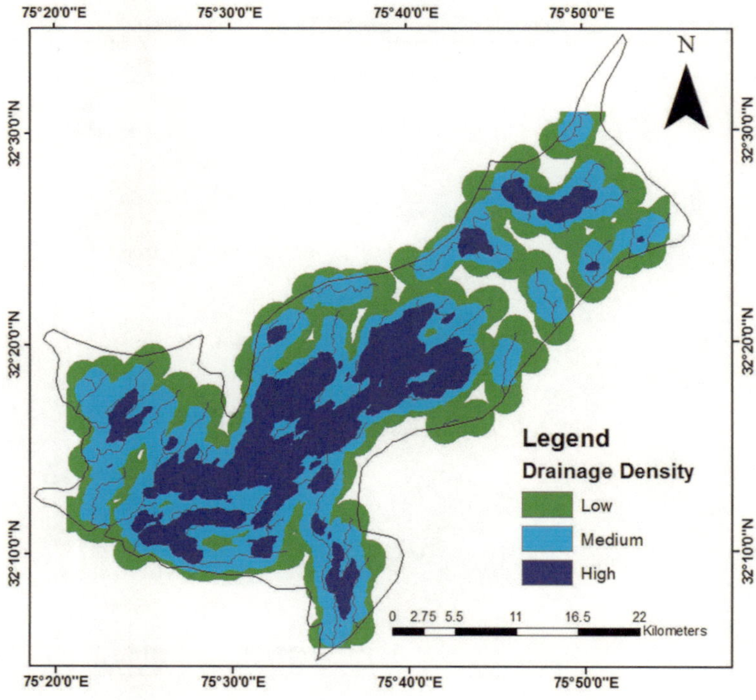

Figure 6.
Drainage map.

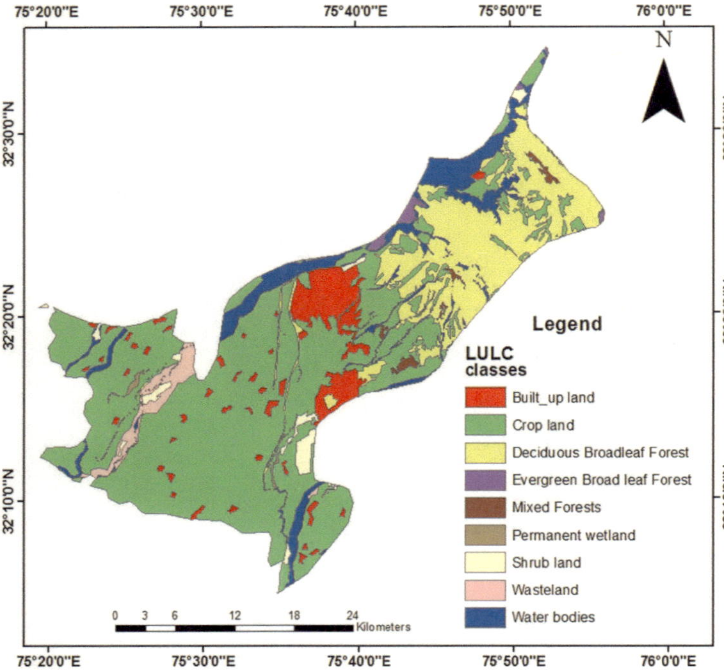

Figure 7.
Land use and land cover.

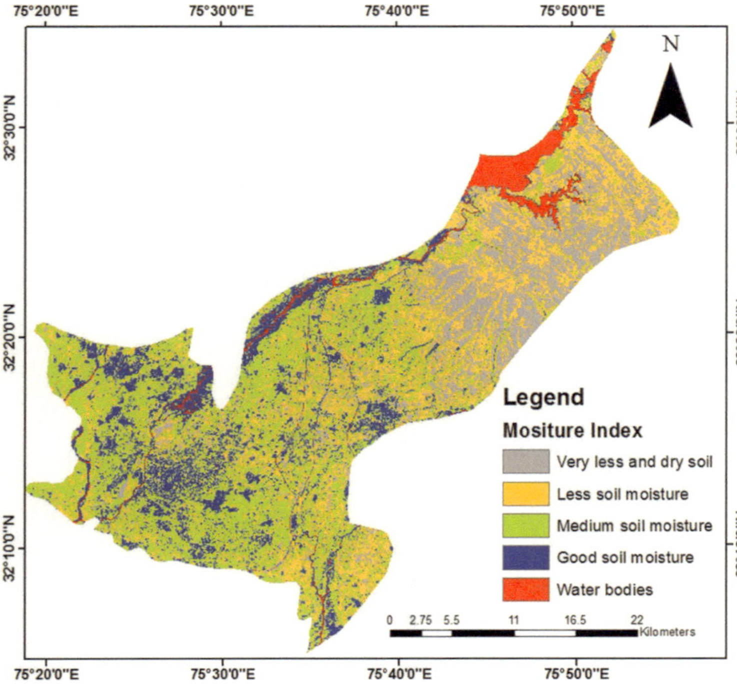

Figure 8.
Moisture index map.

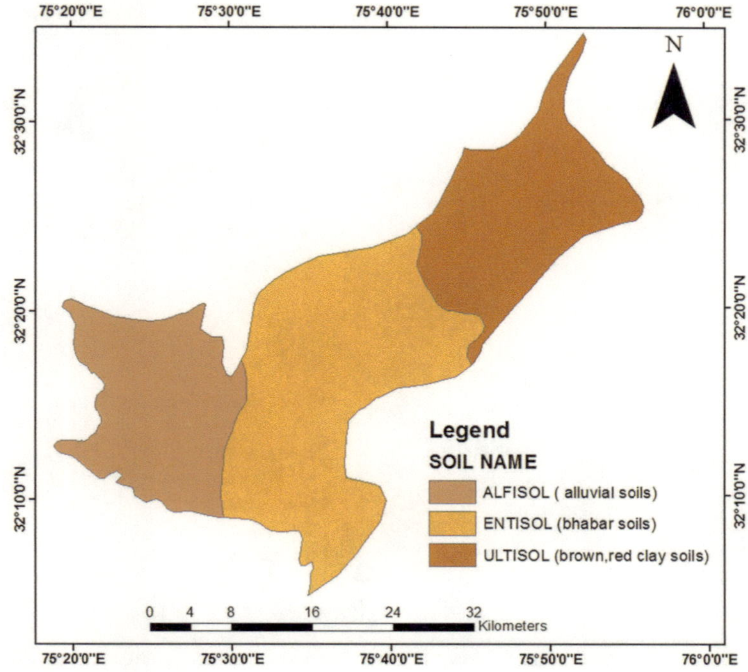

Figure 9.
Soil map.

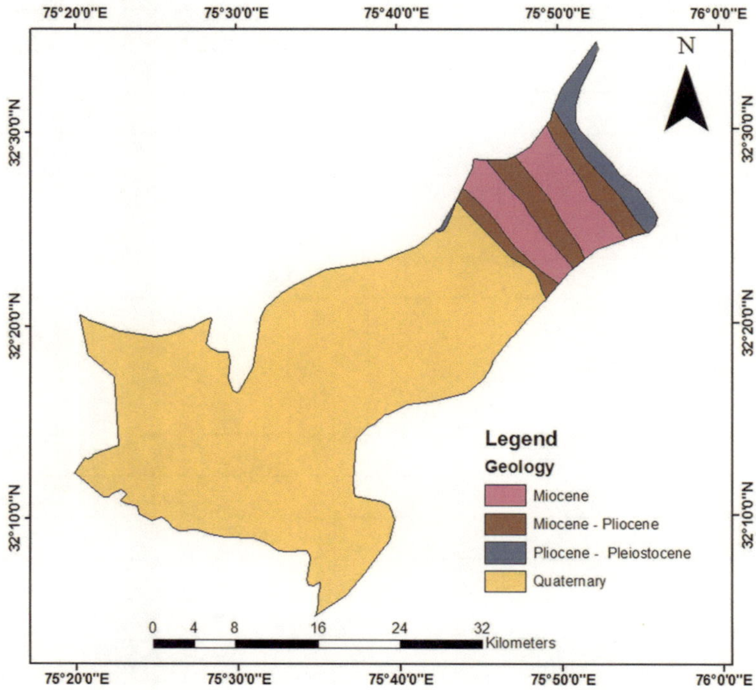

Figure 10.
Geology map.

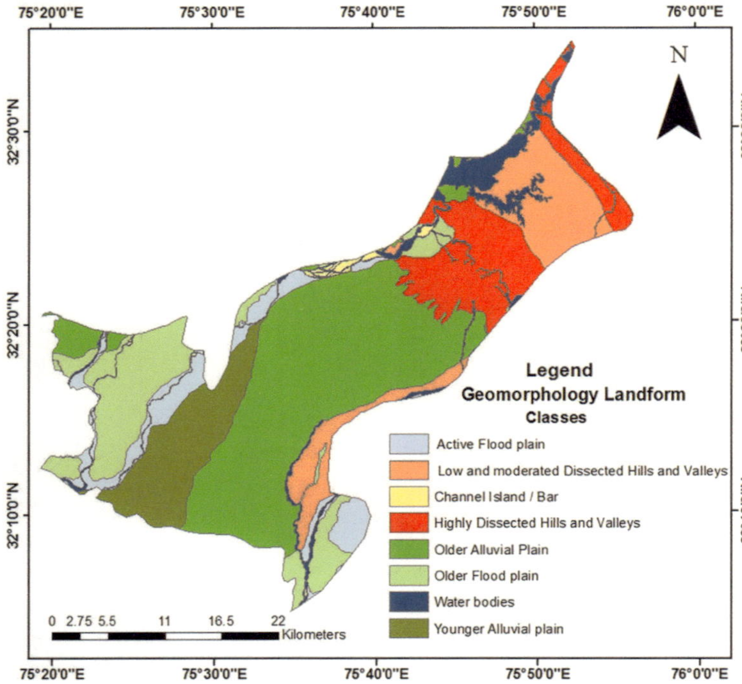

Figure 11.
Geomorphology landform map.

Slope angle (°)	Area (ha)
0–3	66049.4
3–6	13244.1
6–12	8884
12–18	6252.5
18–36	5234.1
36–58	144.8

Table 3.
Slope angle and its area coverage.

Elevation (m)	Area (ha)
224–300	46573.82
300–400	21522.16
400–550	16241.04
550–700	13350.39
700–960	2098.89

Table 4.
Elevation and its area coverage.

Aspect	Area (ha)
East, West	25205.63
North	10275.67
Northeast, Northwest	21756.82
South, Southwest, Southeast	42581.89

Table 5.
Aspect and its area coverage.

Drainage density	Area (ha)
High	24970.4
Medium	37780.1
Low	35291.3

Table 6.
Drainage density and its area coverage.

Class	Area (ha)
Water bodies	8967.44
Evergreen broad leaf forest	595.94
Crop land	57921.61
Built up area	6828.81
Deciduous broadleaf forest	19649.35
Shrub land	1749.07
Permanent wetland	210.61
Wasteland	2236.91
Mixed forest	623.53

Table 7.
LULC and its area coverage.

Moisture index	Area (ha)
Good Soil Moisture	13341.96
Medium Soil Moisture	37086.93
Less Soil Moisture	28455.66
Very Less and Dry Soil Moisture	16474.77
Water Bodies	4453.2

Table 8.
Moisture index and its area coverage.

Soil type	Area (ha)
Alfisol (alluvial soil)	25614.38
Entisol (bhabar soil)	46344.83
Ultisol (brown, red clay soil)	27851.02

Table 9.
Soil and its area coverage.

Geology type	Area (ha)
Miocene	8693.8
Miocene - Pliocene	6611.5
Pliocene - Pleiostocene	3241.69

Table 10.
Geology and its area coverage.

Geomorphology landform	Area (ha)
Active Flood plain	6612
Older Flood plain	12,852
Low and moderated Dissected Hills and Valleys	11,488
Water bodies	5848
Older Alluvial Plain	34,920
Highly Dissected Hills and Valleys	12,364
Channel Island / Bar	572
Younger Alluvial plain	9736

Table 11.
Geomorphology landform and its area coverage.

3. Methodology

Because all the selected criteria are in different units, they must be converted to the same units in order to use the Weighted Overlay Method, which necessitates the use of a standardized value. Standardization techniques transform measurements into uniform units, and the resulting score loses its dimension as well as the unit of measurement for every criterion [21]. All the criteria maps' vector layers were transformed to raster layers. After that, all raster layers were categorized and utilized as input data for the weighted overlay method, which resulted in the creation of the agricultural suitability map. The sub-criteria were ranked on a scale of one to ten, with one being the least significant and ten being the most significant.

One of the most important multicriteria decision-making strategies is the analytical hierarchy process. The procedure is used for a set of criteria or sub-criteria to create a hierarchical structure by assigning weight to each criterion [22].

Relative importance	Definition	Explanation
1	Equal importance	Two criteria enrich equally to the objective criteria
3	Low importance of one over another	Judgments and experience slightly favor one criteria over another
5	Strong or essential importance	Judgments and experience strongly favor
7	Established importance	A criteria is strongly favored and its dominance established in practice
9	Absolute or high importance	The evidence favoring one criteria over another is of the highest probable order of affirmation
2,4,6,8	Intermediate values between the two adjacent importance or judgments	When adjustment is needed

Reciprocals if criteria i *has one of the above numbers designated to it when compared with criteria* j, *then* j *has the reciprocal value when compared with* i.

Table 12.
The fundamental scale for pairwise comparison matrix [25].

The analytic hierarchy process provides a structural foundation for quantifying the strong comparison of design criteria and elements in a paired technique, reducing the decision-making process's complexity [10, 23]. The weight values are determined using a pairwise comparison technique based on the relative significance of the criterion, two at a time [23]. By picking the eigenvalue corresponding to the highest eigenvector of the completed matrix and normalizing the total of the factors to unity, the analytic hierarchy method derives the weights for each individual criterion using the pairwise comparison matrix [4, 24, 25].

The pairwise comparison matrix was generated using the analytic hierarchy procedure described above, using a scale of 1–9, where 9 represents important relevance and 1 indicates equal relevance of the in between criterion of the matrix presented in (**Table 12**) [4, 24, 25].

The reciprocity criteria are mostly used in the comparison matrix, which is mathematically stated as n $(n-1)/2$ for the n number of components in a pairwise comparison matrix [25, 26]. The relative weights and eigenvectors are calculated using Saaty's technique [25] after the pairwise matrix has been computed (**Tables 14** and **15**). Furthermore, one of the major properties of the analytic hierarchy method is that it finds and calculates the inconsistencies of decision makers [24, 25, 27]. The consistency relationship (CR), which is measured by Eq. (1), is used to estimate the efficiency criteria of the analytic hierarchy method.

$$\mathrm{CR} = \mathrm{CI}/\mathrm{RI} \tag{1}$$

The CR is represented by Eq. (1), where CI stands for consistency index and RI stands for random index.

The consistency relationship aids in the determination of possible events and measures the decision maker's/judgments' logical inconsistencies [28–30]. It denotes the probability that the matrix judgments were produced at random [10, 31]. The

n	1	2	3	4	5	6	7	8	9	10
RI	0	0	0.58	0.90	1.12	1.24	1.32	1.41	1.46	1.49

Table 13.
Random inconsistency indices (RI) for n = 10.

Consistency Index and Random Index are the most important factors in determining the CR.

$$CI = (\lambda_{max} - n)/(n - 1) \quad (2)$$

Equation (2) represents the Consistency Index (CI), in which k max is the principle or highest eigenvector of the computed matrix and n is the matrix order.

The Random Index (RI) is the mean value of the consistency index based on the computed matrix order as demonstrated by Saaty [10] (**Table 13**). If the CR value is [0.10], the weight values in the matrix show irregularities, and the approach (AHP) may not produce relevant results [25]. The calculated CR in this investigation was 0.0669, which is within acceptable limits, and the computed weight values are accurate. The obtained weight values are then transformed to percentages in GIS for weighted overlay analysis (WOA), as shown in **Tables 14** and **15** (**Figure 12**).

4. Results and discussion

Weighted Overlay Analysis was carried out to generate the land suitability for agriculture in the Northern region India district using the weight values of selected factors derived from the Analytic Hierarchy Process and specified scores of sub-criteria (**Table 16**). Land suitability for agriculture is classified into five levels, according to the Food and Agricultural Organization (FAO): highly suitable agricultural land, moderately suitable agricultural land, marginally suitable agricultural land, land currently not suitable for agriculture, and permanently not suitable for agricultural production (**Table 17**).

High altitude (224–960 m), high slope (3–58) with higher gully erosion intensity, and less drainage availability of the study area were significant factors, resulting in a smaller area or lower rate of highly appropriate agricultural land in Northern region India (**Figure 13** and **Table 18**).

5. Conclusion

The primary goal of this research was to identify potential agricultural land in the Northern region India district. For the evaluation, an analytical hierarchy approach with a combination of geographic information systems (GIS) was used, and nine different criteria were chosen. The Analytic Hierarchy Process with GIS Integration was shown to be quite useful in determining the best agricultural site. Only 41.2% (39044.28 ha) of the study area was largely suitable for farming at the end of the evaluation, while 40.3% (38139.2 ha) was permanently and temporarily unsuitable for agricultural production. However, inefficient production problems are caused by

Criteria	Slope	Elevation	LULC	Soil moisture	Soil	Geomorphology	Drainage	Geology	Aspect
Slope	1	2	2	3	4	6	7	8	9
Elevation	1/2	1	2	3	4	5	7	7	8
LULC	1/2	1/2	1	4	5	4	6	7	8
Soil moisture	1/3	1/3	1/4	1	3	4	5	6	7
Soil	1/4	1/4	1/5	1/3	1	3	4	5	6
Geomorphology	1/6	1/5	1/4	1/4	1/3	1	3	4	4
Drainage	1/7	1/7	1/6	1/5	1/4	1/3	1	2	3
Geology	1/8	1/7	1/7	1/6	1/5	1/4	1/2	1	3
Aspect	1/9	1/8	1/8	1/7	1/6	1/4	1/3	1/3	1

Table 14.
Pairwise comparison matrix for multi-criteria decision problems.

Criteria	Slope	Elevation	LULC	Soil moisture	Soil	Geomorphology	Drainage	Geology	Aspect	Weights
Slope	0.319	0.426	0.326	0.248	0.223	0.252	0.207	0.198	0.184	0.264
Elevation	0.159	0.213	0.326	0.248	0.223	0.209	0.207	0.173	0.163	0.214
LULC	0.159	0.106	0.163	0.330	0.278	0.168	0.177	0.173	0.163	0.190
Soil moisture	0.105	0.070	0.040	0.083	0.167	0.168	0.148	0.149	.0143	0.119
Soil	0.079	0.053	0.032	0.027	0.056	0.126	0.118	0.124	0.122	0.082
Geomorphology	0.054	0.043	0.040	0.020	0.018	0.042	0.089	0.099	0.081	0.054
Drainage	0.045	0.030	0.027	0.016	0.014	0.014	0.029	0.049	0.061	0.034
Geology	0.039	0.030	0.023	0.014	0.011	0.010	0.015	0.025	0.061	0.025
Aspect	0.035	0.027	0.020	0.012	0.009	0.010	0.009	0.008	0.020	0.018

Maximum eigenvalue (λ_{max}) *= 9.79.*
n = 9.
Consistency index (CI) = (λ_{max} - n)/ (n - 1) *= 0.098.*
Random index (RI) = 1.46.
Consistency ratio (CR) = (CI/RI) = 0.0676 < 0.10.

Table 15.
Normalized pairwise comparison matrix for multi-criteria decision making.

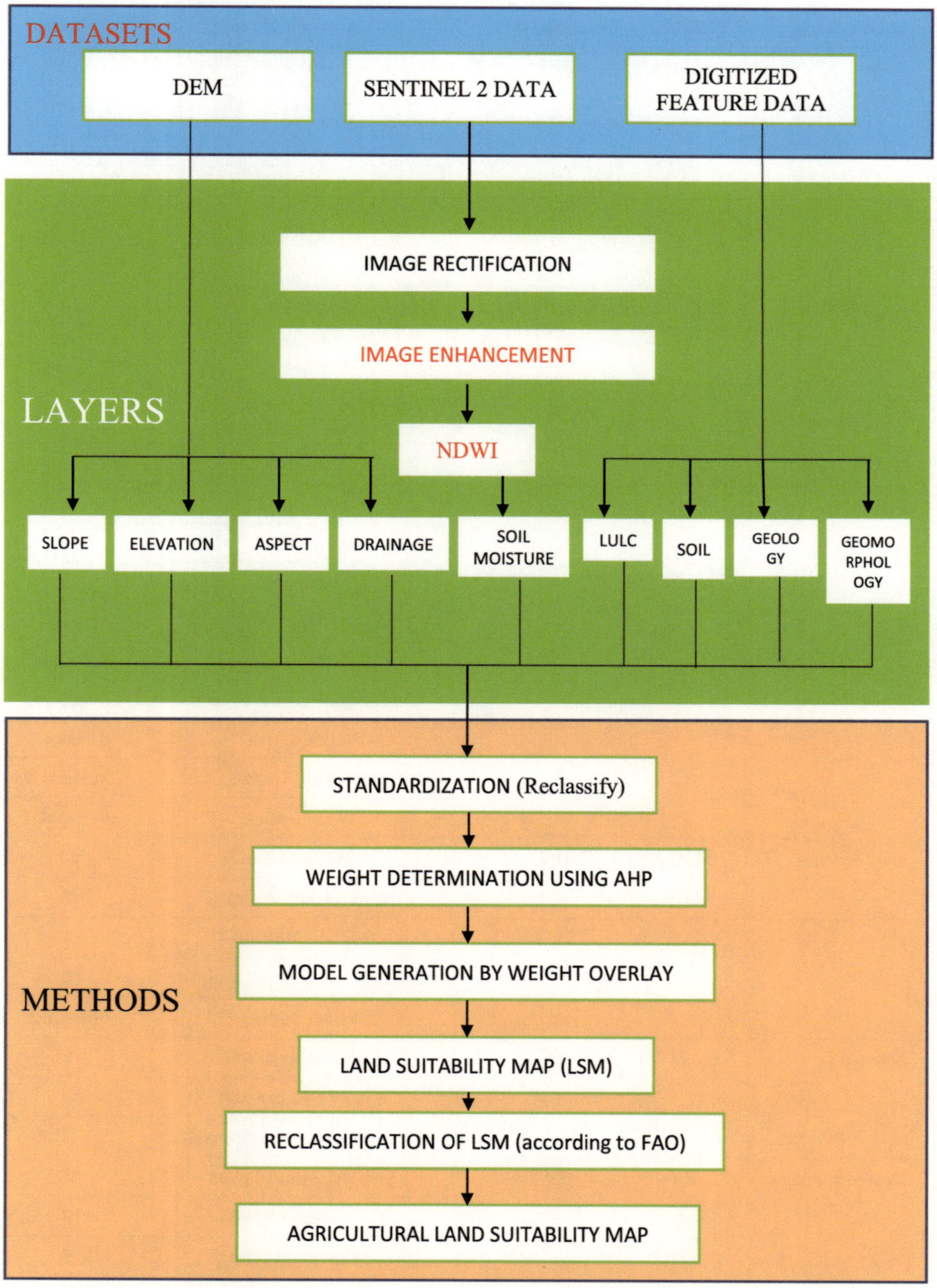

Figure 12.
Procedure followed in generating agricultural land use suitability map.

geomorphological qualities such as very high elevation, steep slope, reduced soil moisture, the presence of bare rocks, and a lack of irrigation system availability. As a result of all these concerns, a moderate quantity of land in the study district has been identified as appropriate for agricultural production. The established result can be implemented into the agricultural production decision-making process in the study

Main criteria	Weight	Influence (%)	Sub-criteria	Score
Slope	0.264	26.4	0–3	10
			3–6	8
			6–12	6
			12–18	4
			18–36	2
			36–58	1
Elevation	0.213	21.4	224–300	10
			300–400	9
			400–550	8
			550–700	7
			700–960	5
LULC	0.190	19	Crop Land	10
			Shrub Land	4
			Wasteland	3
			Evergreen Broad leaf Forest	Restricted
			Mixed Forest	Restricted
			Deciduous Broadleaf Forest	Restricted
			Built up Area	Restricted
			Permanent wetland	Restricted
			Water bodies	Restricted
Soil Moisture	0.119	11.9	Good Soil Moisture	10
			Medium Soil Moisture	7
			Less Soil Moisture	4
			Very Less and Dry Soil Moisture	1
			Water Bodies	Restricted
Soil	0.082	8.2	Alfisol (alluvial soil)	9
			Entisol (bhabar soil)	6
			Ultisol (brown, red clay soil)	3
Geomorphology	0.054	5.4	Younger Alluvial plain	10
			Older Alluvial Plain	9
			Older Flood plain	8
			Low and moderated Dissected Hills	3
			Active Flood plain	2
			Channel Island / Bar	1
			Highly Dissected Hills and Valleys	1
			Water bodies	Restricted

Main criteria	Weight	Influence (%)	Sub-criteria	Score
Drainage	0.034	3.4	High	9
			Medium	7
			Low	4
Geology	0.025	2.5	Quaternary	9
			Pliocene - Pleiostocene	7
			Miocene - Pliocene	5
			Miocene	3
Aspect	0.018	1.8	South, Southwest, Southeast	9
			East, West	5
			Northeast, Northwest	4
			North	2

Table 16.
Weights of the criteria and scores of the sub-criteria.

Suitability level	Suitable areas for agricultural production	
	Area(ha)	%
High suitability	390442.28	41.2
Moderate suitability	13498.76	14.3
Marginally suitable	3993	4.2
Currently not suitable	1766.6	1.9
Permanently not suitable	36372.6	38.4

Table 17.
Areal and percentile distribution of agricultural land suitability analysis results.

area, as it provides insight into determining suitable sites. By critically assessing the procedures and approaches used, the results can be more precise. Physical elements (topographical properties, soil and geological characteristics, etc.) are only part of the analysis, which must also include economic and social conditions for agricultural production. Because the pairwise comparison approach is based on expert judgments, which are primarily subjective in nature, it is used in the analytic hierarchy process. As a result, any incorrect judgment on any of the selected factors can be effectively communicated to the score assignment and weight designation. This is the main disadvantage of the analytic hierarchy approach; hence, weights and scores must be carefully chosen [32, 33]. For more helpful and accurate results, the study should focus on a few key species, such as several therapeutic plants and species that have substantial economic worth and also influence the advancement of rural tourism. The use of very high-resolution satellite images will aid in the assessment of finer areas.

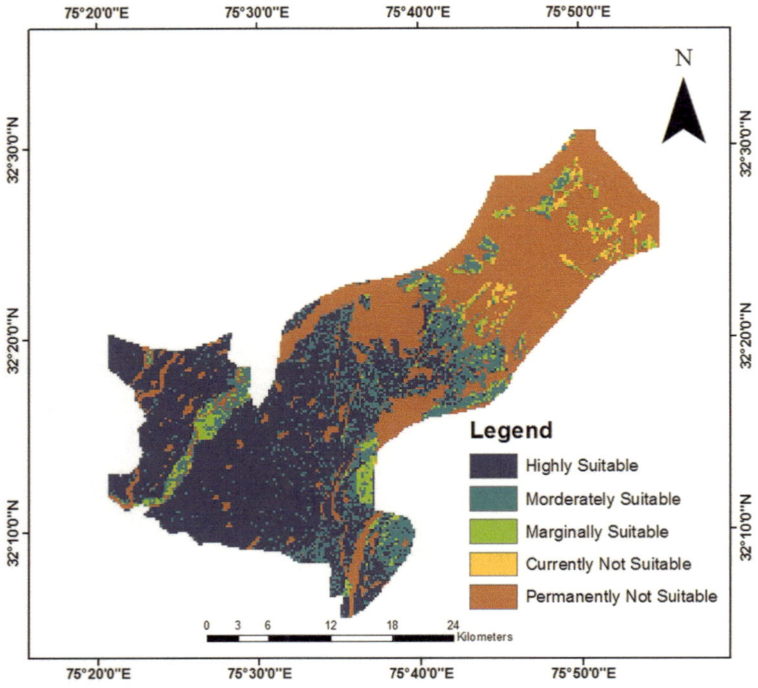

Figure 13.
Agriculture land suitability map of northern region India.

Suitability level	Suitable areas for agricultural production		Land qualities/characteristics	Remarks
	Area (ha)	**%**		
High suitability	390442.28	41.2	Gentle slopes (0–3) with gullies, high soil moisture with lower elevation, alluvial soil, good drainage capacity	Most suitable for agriculture, favorable area for intensive agriculture if irrigation facilities are available
Moderate suitability	13498.76	14.3	Gentle to stiff slopes (3–10) with micro terracing, medium soil moisture with lower elevation, moderate drainage capacity	Suitable land for farming practices with proper management, suitable for terrace cultivation
Marginally suitable	3993	4.2	(10–20) slope, less soil moisture with higher elevation, coarse loamy to gravel loamy soil, low drainage availability	Less suitable land for agriculture with careful farm management, necessary protections from drainage and intensive erosion
Currently and permanently not suitable	38139.2	40.3	Precipitous slope with rocky lands, dry soil, dense forest, barren land, loamy skeletal soil, no drainage availability	The land is not suitable for agriculture, areas under dense vegetation, settlement, barren lands, open rocks are not suitable for agriculture

Table 18.
Land suitability levels and their land characteristics.

Before the ultimate implementation, the indicated locations must also be documented on the ground with various other local and regional parameters.

Acknowledgements

Authors are very much grateful of the Director, Defense Research & Development Organization, (DRDO) Govt. of India, New Delhi, for providing infrastructure and facilities during the work was carried out.

Author details

Mujahid Ali Khan, Rizwan Ahmad* and Haris Hasan Khan
Interdisciplinary Department of Remote Sensing and GIS Applications, Aligarh Muslim University, Aligarh, India

*Address all correspondence to: rahmed1378@gmail.com

References

[1] Beek KJ, Burrough PA. Quantified Land Evaluation Procedures: Proceedings of the International Workshop. Enschede, Netherlands: International Institute for Aerospace Survey and Earth Sciences; 1987

[2] de la Rosa D. Micro LEIS 2000: Conceptual Framework. Instituto de Recursos naturales y Agrobiologia, CSIC, Avda: Sevilla, Spain; 2000

[3] Rossiter DG. A theoratical framework for land evaluation. Geoderma. 1996;**72**: 165-202

[4] Malczewski J. GIS and Multicriteria Decision Analysis. USA and Canada: John Wiley & Sons; 1999

[5] Tkach RJ, Simonovic SP. A new approach to multi-criteria decision making in water resources. Journal of Geographic Information and Decision Analysis. 1997;**1**(1):25-43

[6] Jankowski P. Integrating geographical information systems and multiple criteria decision making methods. International Journal of Geographic Information System. 1995;**3**(2):251-273

[7] Pereira JMC, Duckstein L. A multiple criteria decision-making approach to GIS- based land suitability evaluation. International Journal of Geographical Information Science. 1993; 7(5):407-424

[8] Ishizaka A, Labib A. Review of the main developments in the analytic hierarchy process. Expert Systems with Applications. 2011;**38**(11):14336-14345

[9] Saaty T. An Eigenvalue Allocation Model for Prioritization and Planning. In Working Paper. Pennsylvania: Energy Management and Policy Center University of Pennsylvania; 1972

[10] Saaty TL. A scaling method for priorities in hierarchical structure. Journal of Mathematical Psychology. 1977;**15**:3

[11] Shang J, Sueyoshi T. A unified framework for the selection of a flexible manufacturing system. European Journal of Operational Research. 1995;**85**(2): 297-315

[12] Myint S, Tabucanon MT. A multiple-criteria approach to machine selection for flexible manufacturing systems. International Journal of Production Economics. 1994;**33**(1–3):121-131

[13] Liberatore MJ. An extension of the analytic hierarchy process for industrial R&D project selection and resource allocation. IEEE Transactions on Engineering Management. 1987;**1**:12-18

[14] Khorramshahgol R, Moustakis VS. Delphic hierarchy process (DHP): A methodology for priority setting derived from the Delphi method and analytical hierarchy process. European Journal of Operational Research. 1988;**37**(3): 347-354

[15] Duran O, Aguilo J. Computer-aided machine-tool selection based on a fuzzy-AHP approach. Expert Systems with Applications. 2008;**34**(3):1787-1794

[16] Ayağ Z, Özdemir RG. A fuzzy AHP approach to evaluating machine tool alternatives. Journal of Intelligent Manufacturing. 2006;**17**(2):179-190

[17] Chang CW, Wu CR, Lin HL. Integrating fuzzy theory and hierarchy concepts to evaluate software quality.

Software Quality Journal. 2008;**16**(2): 263-276

[18] Xu L, Li Z, Li S, Tang F. A decision support system for product design in concurrent engineering. Decision Support Systems. 2007;**42**(4): 2029-2042

[19] Lin YJ, Huang CW, Tseng JC, Shiau JY. Issue resolution for conceptual design using AHP. In: Intelligent Systems in Design and Manufacturing. Vol. 5605. Bellingham: SPIE; 2004. pp. 47-53

[20] Adhikari I, Kim SY, Lee YD. Selection of appropriate schedule delay analysis method: Analytical hierarchy process (AHP). In: Proceedings on PICMET. IEEE: Istanbul, Turkey; 2006

[21] Effat HA, Hassan OA. Designing and evaluation of three alternatives highway routes using the analytical hierarchy process and the least-cost path analysis, application in Sinai peninsula, Egypt. Egyptian Journal of Remote Sensing and Space Science. 2013;**16**(2):141-151

[22] Kiker GA, Bridges TS, Varghese A, Seager TP, Linkov I. Application of multicriteria decision analysis in environmental decision making. Integrated Environmental Assessment and Management. 2005;**2**:95-108

[23] Miller W, Collins W, Steiner FR, Cook E. An approach for greenway suitability analysis landscape and urban planning. International Journal of Geographical Information Science. 1998; **42**(2–4):91-105

[24] Feizizadeh B, Jankowski P, Blaschke T. A GIS based spatially-explicit sensitivity and uncertainty analysis approach for multi-criteria decision analysis. Computational Geosciences. 2014;**64**:81-95

[25] Thomas PG, Doherty PC. The analytic hierarchy. In: Process: Planning Priority Setting, Resource Allocation. New York: McGraw-Hill; 1980

[26] Akinci H, Ozalp AY, Turgut B. Agriculture land use suitability analysis using GIS and AHP technique. Computers and Electronics in Agriculture. 2013;**97**:71-82

[27] Garcia JL, Alvarado A, Blanco J, Jimenez E, Maldonado AA, Cortés G. Multi-attribute evaluation and selection of sites for agricultural product warehouses based on an analytic hierarchy process. Computers and Electronics in Agriculture. 2014;**100**: 60-69

[28] Cengiz T, Akbulak C. Application of analytical hierarchy process and geographic information systems in land-use suitability evaluation: A case study of Dumrek village. The International Journal of Sustainable Development & World Ecology. 2009;**16**(4):286-294

[29] Chen H, Liu G, Yang Y, Ye X, Shi Z. Comprehensive evaluation of tobacco ecological suitability of Henan Province based on GIS. Agricultural Sciences in China. 2010a;**9**(4):583-592

[30] Chen Y, Yu J, Khan S. Spatial sensitivity analysis of multi-criteria weights in GIS-based land suitability evaluation. Environmental Modelling & Software. 2010;**25**(12):1582-1591

[31] Park S, Jeon S, Kim S, Choi C. Prediction and comparison of urban growth by land suitability index mapping using GIS and RS in South Korea. Landscape and Urban Planning. 2011;**99**(2):104-114

[32] Kritikos TRH, Davies TRH. GIS-based multi-criteria decision analysis for landslide susceptibility mapping at

northern Evia, Greece [GIS-basierte multikriterielle Entscheidungsanalysen zur Kartierung von Massenverlagerungspotenzialen im nördlichen Evia. Griechenl]. Z. Dtsch. Ges. Für Geowiss. 2011;**162**:421-434

[33] Nefeslioglu HA, Sezer EA, Gokceoglu C, Ayas Z. A modified analytical hierarchy process (M-AHP) approach for decision support systems in natural hazard assessments. Comput. Geosci. 2013;**59**:1-8

Section 5

Solid Waste Management

Chapter 7

Application of Geographic Information System in Solid Waste Management

Elsai Mati Asefa, Kefelegn Bayu Barasa and Dechasa Adare Mengistu

Abstract

The application of geographic information systems (GIS) to solid waste management (SWM) has been widely adopted in many cities around the world. Planning a sustainable waste management approach is complex, tedious, and time-consuming, and decision-makers are frequently subjected to conflicting factors. GIS has a crucial role in simplifying and facilitating the implementation of sustainable SWM. It is a powerful tool that can assist in minimizing value conflicts among preference and interest parties by providing better information. In this chapter, the basic principles of how GIS is utilized in SWM planning are discussed. The first few sections deal with sustainable SWM planning, its challenges, and problems with the poor performance of its planning. Furthermore, the principles of GIS, how it evolved in SWM, and its integration with multi-criteria evaluation were discussed. The final sections deal with the application of GIS in waste collection optimization and waste disposal planning. The primary aim of this chapter is, therefore, to aid decision-makers in the field so that they can apply it to the daily challenges of SWM.

Keywords: waste management, solid waste, GIS, MCDA, route planning, landfill, spatial analysis

1. Introduction

In recent decades, the application of geographic information systems (GIS) in solid waste management (SWM) has been widely adopted in many cities around the world, from developed to developing nations. Moreover, the utilization of GIS is not limited to the management of solid waste. It has been widely applied in agriculture, natural resource management, planning and economic development, disaster management and mitigation, public health, and related areas. In solid waste management, the primary objective for the adoption of GIS is to reduce cost and time (feasibility) and also to help planners make better decisions in designing solid waste management.

Essentially, planning a sustainable waste management approach is complex, tedious, and time-consuming, and decision-makers are frequently subjected to conflicting factors in SWM planning. Sustainable waste management, according to the United Nations, is aimed at the integration of SWM at the national and local level, utilizing a life-cycle approach for resource efficiency and environmentally safe management of solid waste [1]. Waste reduction, resource reuse, recycling, and recovery are all part of sustainable waste management, which helps to reduce pollution while also extending the life of resources to be wasted. As a result, sustainable waste management should be economically feasible, socially acceptable, and environmentally effective upon implementation.

The geographic information system can be used to simplify the ease of implementation of sustainable SWM. Most often, due to the complex nature of SWM, there is poor performance in its technical functions and management functions (policy and legislation, financial, and stakeholders). Owing to this fact, there are different optimization techniques developed and implemented in different areas. Among the techniques used for spatial and non-spatial data information, GIS is common. In most cases, GIS is used with the integration of remote sensing and/or multicriteria decision-making analysis (MCDA). In spatial multicriteria analysis, the combination of GIS and MCDA capabilities is crucial. GIS allows for the acquisition, storage, retrieval, manipulation, and analysis of data to obtain information for decision-making. Whereas MCDA techniques provide tools for aggregating geographic data and decision-maker preferences into a single-dimensional value or utility of alternative decisions [2].

A brief description of how GIS is used in solid waste management was presented in this chapter. Our main goal was to show the fundamentals of GIS in SWM to individuals who have little or no experience with SWM planning, especially in developing countries. Furthermore, experts who specialized in GIS or SWM and have more experience with one than the other should understand the fundamentals of integrating the two to maximize efficiency. In addition, this chapter covers the concepts of sustainable waste management, GIS-based SWM, and MCDA. The application of GIS in solid waste collection optimization and waste disposal planning is discussed also.

2. Concepts of sustainable solid waste management

2.1 Solid waste management

Solid waste is defined as any waste discarded by households, commercial, mining, and agricultural operations, as well as the residue sludge from wastewater treatment plants, water supply treatment plants, and air pollution control centers, and it is other than liquids and gases [3]. Solid wastes can be categorized based on sources, reusable potential, degree of biodegradability, and potential impact on the environment, as shown in **Table 1**. Solid waste management (SWM) is a discipline related to the proper management of solid waste generation, storage, collection, transfer, transport, processing, and disposal of solid waste [5].

Solid waste management (SWM) consists of six functional elements, which include solid waste generation, collection, transfer, storage, processing, and disposal. It is a discipline related to the proper management of each of these functional elements. Waste collection is the process of gathering solid waste and recyclable materials and transporting them to a destination where the collection vehicle is emptied, such as a materials-processing facility, a transfer station, or a landfill, once they have

Classification	Type	Description	Composition
Source (based on [4])	Residential	From Single and multifamily dwellings	Food wastes, paper, cardboard, plastics, textiles, leather, yard wastes, wood, glass, metals, ashes, special wastes (e.g., bulky items, consumer electronics, white goods, batteries, oil, tires), and household hazardous wastes.).
	Commercial	From stores, hotels, restaurants, markets, office buildings, etc.	Paper, cardboard, plastics, wood, food wastes, glass, metals, special wastes, hazardous wastes.
	Industrial	From light and heavy manufacturing, fabrication, construction sites, power, and chemical plants.	Housekeeping wastes, packaging, food wastes, construction and demolition materials, hazardous wastes, ashes, special wastes.
	Institutional	From schools, hospitals, prisons, government centers.	Same as the commercial wastes
	Construction and demolition	New construction sites, road repair, renovation sites, demolition of buildings	Wood, steel, concrete, dirt, etc.
	Agriculture	Crops, orchards, vineyards, dairies, feedlots, farms.	Spoiled food wastes, agricultural wastes, hazardous wastes (e.g., pesticides).
Degree of biodegradability	Biodegradable	Can be decomposed easily by bacteria or any other natural organisms	Food wastes, garden wastes, paper, cardboard
	Nonbiodegradable	Cannot be decomposed or degraded by the biological process	Plastics
Potential impact	Hazardous	Wastes whose uses or disposal pose a threat to human health or the environment	Toxic, corrosive explosive, and/ or inflammable Pesticides, herbicides, paints, industrial solvents, fluorescent light bulbs, and mercury-containing batteries
	Nonhazardous	Wastes that are considered less harmful to the environment or human health	Paper, plastics, glass, metals

Table 1.
The categories and classification of solid wastes.

been collected [6]. Following the collection of these wastes, the wastes are sorted, processed, and reused to recover, recycle, or reuse them. Additionally, the waste could be treated to recover energy. After the recovery of products, the majority of the waste is compacted to reduce volume, weight, and size. Finally, the remaining waste is either disposed of through regulated methods such as landfilling or thrown in the open.

2.2 Integrated solid waste management and zero waste principle

Integrated solid waste management (ISWM) is a strategic approach to sustainable solid waste management in which solid waste generation, segregation, transportation, sorting, treatment, recovery, and disposal are all combined with the goal of resource efficiency. The US Environmental Protection Agency defines ISWM as the prevention, recycling, composting, and disposal of solid waste to protect public health and the environment, with a focus on reducing recycling, and managing waste [7]. Solid waste management that is integrated can provide both environmental and economic sustainability. No single waste management technology can deal with all waste products in an environmentally sustainable manner. A wide range of management options is ideal. As a result, any waste management system is made up of several interconnected processes. This method examines the entire waste management system and offers methods for estimating overall environmental and economic costs [8].

To bring about a sustainable world for present and future generations, besides the ISWM, there are many approaches to SWM today, including the "zero waste" approach and the life cycle inventory approach. The "Zero waste" approach has been defined in various ways by different individuals. Literally, zero waste is defined as the complete elimination and absence of waste. However, the generation of waste is inevitable as it is a result of extraction and manufacture, distribution, consumption, and other daily activities of human beings. As per the US Environmental Protection Agency, "zero waste" is defined as "the conservation of all resources through responsible production, consumption, reuse, and recovery of products, packaging, and materials without burning or discharges to land, water, or air that endanger the environment or human health". It is an approach that aims to optimize waste recycling and reduction; products are designed in a manner that can be reused, mended, or recycled back into nature [9].

2.3 Sustainable solid waste management

The collection, transportation, treatment, and disposal of diverse types of waste in a manner that does not damage the environment, human health, or future generations is referred to as sustainable waste management. It encompasses all aspects of waste management organization, from production to final disposal [10]. To minimize irreversible negative impacts on human health and the environment, the sustainability concept suggests that industries and their operations be encouraged to become more efficient in terms of resource utilization, production of less pollution and process waste, and use of nonrenewable resources [11]. The waste management hierarchy, on the other hand, frequently necessitates more effort, invention, and creativity, as well as good regulations, stakeholders, and financial support, to achieve the "zero waste" goal.

2.4 Problems with solid waste management

2.4.1 Impacts of solid waste management

Solid waste management (SWM) has become a tough task for many cities around the world due to a significant increase in waste generation as a result of demo-technic growth, population growth combined with technological advancement [12, 13]. Waste collection, treatment, and disposal, among other issues, are crucial due to their costs [13]. Poor SW management has serious environmental implications and puts public health at risk. Moreover, the problem is worse in developing countries that lack basic infrastructure and are known for their less organized waste management practices coupled with a high waste generation rate [14]. According to Tan [12], in these countries, the collection rate is as low as 30–50%, and even the collected waste ends up in unmanaged landfills. Another major challenge is inadequate policy and legislation, lack of public commitment and awareness, lack of technical capacity, and poor financing in these countries [15].

Furthermore, gas and leachate generation are unavoidable outcomes of solid waste disposal in landfills, owing to microbial decomposition, climatic circumstances, refuse properties, and landfilling processes. The migration of gas and leachate away from landfill limits and their release into the surrounding ecosystem pose major environmental concerns at both existing and new sites. These concerns include but are not limited to, fires and explosions, vegetation damage, foul odors, landfill settlement, groundwater pollution, air pollution, and global warming, among others [16].

3. A new way of thinking

Recently, the implementation of integrated solid waste management (ISWM) has become popular for better management of growing MSW. It ideally proposes the waste reduction at source, before even generation of wastes utilizing different techniques, innovations, and optimal management practices [17]. In ISWM, there is effective management of waste at all levels, from generation to disposal. Despite its popularity, little is known about its technical operation, and it is also subjective from a management perspective.

Integrated solid waste management compromises the complex multi-objective criteria in each management element. This resulted in making the ISWM more subjective for decision-makers as uncertain solid waste generation, conflicting factors, and economic constraints are dealt with. To overcome these problems, there are some innovations in ISWM, including mathematical modeling, computer-based modeling, geographic information systems, and remote sensing, to name a few. Besides, Gaeta et al. [18] divided the innovations in ISWM grouping into four main typologies of innovations in the solid waste market system as follows: a traditional landfill-oriented system; a modern waste-to-energy incinerator-oriented system; a light recycling system; and a hard recycling system. In this chapter, the focus is on GIS, also detailed discussions on other approaches are presented by [14, 18–21].

To optimize solid waste management, there were some early attempts. For example, Anderson & Nigam [22] were the first to propose mathematical modeling in SWM. Followed by the development of models for technology selection, siting, and sizing of waste processing facilities in SWM [23–25]. On the other hand, geographic information systems combined with remote sensing are becoming widely common in the area of SWM. GIS is preferred because of its simplicity, easy access, and low cost [26]. The use of GIS in solid waste management supports capturing, handling, and transmitting the required information promptly and properly, and it is a well-known, innovative technology that has contributed a lot to SWM in a short period.

4. GIS and SWM

4.1 History and evolution of GIS in SWM

Historically, early society relied heavily on spatial data to represent geographical locations. In the American War of Independence, at the Battle of Yorktown, the French cartographer prepared the map overlays of troop movements [27]. Also, in 1854, Dr. John Snow, in providing evidence for a water-borne cholera outbreak, mapped the incidence of cases with water supply sources. Even though there was a huge reliance on spatial data in early societies, the breakthrough was witnessed with the introduction of GIS after a century. The precise articulation of GIS's brief history is muddled at best. This is because most organizations involved in the use of GIS at the time of its inception refused to give up their data, and the early writers of GIS history were not practical users of the technology [28]. Even though the book did not give any description of the overlay processing with defined methodology, the publication of a book named "Design with Nature" by McHarg [29] is known to have been a crucial factor in the evolution of GIS. Many research articles, reviews, and books concur that the evolution of GIS over time is due to improvements in geography, environmental awareness, technology advancements, and the enhanced practical and technical skills of GIS users.

Geographic information arose as integrating and powerful technology in the context of these breakthroughs because it allowed researchers and geographers to include their methods and information in a variety of ways that supported traditional kinds of geographic analysis. For example, map overlay analysis and other forms of analysis and modeling that were previously impossible to achieve using manual approaches [30]. Researchers can now map, model, query, and analyze large volumes of data in a single database thanks to GIS [31]. GIS can play a part in SWM because it is complicated and the components are interconnected. As a result, the planning and monitoring operations are based on spatial data. Customer service, analyzing optimal transfer station locations, planning routes for vehicles transporting waste from residential, commercial, and industrial customers to transfer stations and from transfer stations to landfills, locating new landfills, and monitoring the landfill are all important aspects of SWM that GIS can help with.

4.2 Role of GIS in SWM

In the management of solid waste, a geographic information system is an excellent instrument as it is used in the planning of technical elements. Many researchers have used GIS principles in SWM to optimize the practice of SWM. For example, Chang et al. [32] used GIS in conjunction with a mathematical programming model to develop a multi-objective, mixed-integer programming model for collection vehicle routing and scheduling for solid waste management systems synthesized within a GIS environment and concluded the effectiveness of the model, as well as recommendations for application to other environmental planning and management problems. In recent decades, efforts have been undertaken to shorten the distance between waste collection stations and landfills, hence reducing the number of trucks involved in waste collection and disposal [33–36].

GIS can be used to save costs and improve waste collection and transportation efficiency. Many elements influence route optimization, including the location of waste bins, collection details, vehicle kinds, trip impedances, and the road network's integrity [37]. Chang and Lin [38] used a GIS and a mixed-integer programming model to locate proper waste bin locations and waste transfer stations, which resulted in lower direct costs and more manageable operational programs. Even though the process varies, the technique has recently become popular. El-Hallaq & Mosabeh [39] used the GIS integrated location-allocation methodology to rebuild the existing waste bin sites and were successful in finding misplaced bins and recommending an equitable waste bin distribution.

Another key application of GIS in SWM is the selection of disposal sites. Muttiah et al. [40] used a GIS and a Markov-chain-based simulated annealing algorithm to find prospective waste disposal sites, and the simulated annealing method saved order of magnitude of time over an exhaustive search strategy. Multi-criteria coupled with GIS have recently become popular in SWM for assessing conflicting criteria.

Stages	Description	Principles/basics	Examples
Input	Identifying and gathering data related to SWM. Acquisition, reformatting, georeferencing, compiling and documenting these data.	Digitizing, scanning, remote sensing, GPS, internet	Daily waste generation rate, waste collection routes, landfill site location, type of waste, land use map, the elevation of the area.
Data storage and management	Includes those functions needed to store and retrieve data from the database can be thought of as a representation or model of real-world geographical systems	Geographical entity (towns, road network, and town boundary represented by point, line, and polygon). Object (spatial and nonspatial data)	Data model location of the existing landfill site (spatial) and types of waste dumped with the rate of dumping (nonspatial data).
Data manipulation and storage	To obtain information useful for waste collection optimization or waste disposal planning.	Fundamental analysis (measurement, classification, overlay operations, and neighborhood and connectivity operations) and advanced analysis (statistical modeling and mathematical modeling)	Vehicle route optimization, suitability analysis of landfill, optimal waste bin location
Data output	A way to see the analyzed SWM-related data or information in the form of maps, tables, diagrams.	Display monitors, pen plotters, electrostatic plotters, laser printers, line printers, and dot matrix printers and plotters	Optimal route for waste collection (map), schedule for waste collection (table), thematic map of a suitable landfill site.

SWM: solid waste management; GPS: global positioning service.

Table 2.
Basic principles of GIS.

Asefa et al. [41] used GIS in combination with the multicriteria decision-making method and the analytical hierarchy process to find the best landfill location by balancing competing environmental and socioeconomic concerns. In addition, Rahimi et al. [42] used GIS techniques and fuzzy Multi-Criteria Decision-Making methods in landfill site selection problems, where the criteria weights are determined using the group fuzzy Best-Worst Method, suitability maps are generated using GIS analysis, and the sites are analyzed and ranked using the group fuzzy MULTIMOORA method.

4.3 Basic principles of GIS

A geographic information system (GIS) has a broad area of application and is a computer-based system that helps in the manipulation of data about specific geographic areas. GIS, as presented by Rolf and Deby [43], is a georeferenced data entry, analysis, and presentation tool based on the computer interface. Data preparation and entry is the initial phase in GIS processing, which entails acquiring, preparing, and entering the data needed for information production into the GIS database system. The second stage is data analysis, which entails going over and analyzing the information that has been gathered and uploaded into the GIS system. Finally, during the data presentation phase, the analytical results are displayed and/or saved appropriately. Data input, data management, analysis, and final output are usually common processes in most GIS technique development for various purposes and applications, depending on the basic concept of the above-mentioned stages. **Table 2** summarizes and presents these stages, as well as their fundamental principles and instances.

5. Geospatial analysis

Geospatial analysis is the process of calculating data that has been entered or saved to generate new information that can be used to improve SWM decisions. The decision on which geographical analysis to use is based on the decision maker's needs and objectives. For example, when deciding where to build a new landfill, various conflicting environmental, social, and political variables must be considered. When these criteria are entered into the GIS interface, other geographical analyses, such as reclassification, overlaying, buffering, and so on, can be performed. As a result, GIS can assist in the computation of such cases using up-to-date criteria that are examined cost-effectively. The following are some of the most often used spatial analysis methodologies as discussed in [2, 43, 44].

6. Measurement, retrieval, and classification

It is usually used at the beginning of any analysis as it allows data exploration without making a significant change. In the measurement function, distances between features or along their perimeters, the counting frequency of features, and the computation of area size features are computed. The retrieval functions allow the selection of specific features based on logical functions and user preferences. Classification is the re-arrangement of specific features into a common data value layer. Usually, measurement, retrieval, and classification are performed using a single vector or raster data layer combined with non-spatial data, sometimes.

7. Overlay functions

The overlay capability, which allows disparate data layers to be joined to create new data, is considered the most important feature of GIS interfaces. It can be used for both vector and raster data types, however, it is most commonly employed for raster overlay computations. It includes operations of intersection, union, difference, and complement using sets of positions. Many GISs allow overlays using an algebraic language, which expresses an overlay function as a formula with data layers as parameters. Arithmetic, relational, and conditional operators, as well as a variety of functions, can be used to combine different layers [2].

8. Neighborhood functions and Network analysis

Neighborhood functions allow the evaluation of the surrounding areas of the location of the features and operate on the neighboring features of a given feature or set of features. It includes search functions to allow the retrieval of features, line-in-polygon and point-in-polygon functions to compute a given linear or point feature is located within a given polygon, and buffering functions which allow determining a fixed-width environment surrounding a feature. Network analysis is concerned with network computations in the GIS interface. A network is a series of interconnected lines that depict a geographic phenomenon, most commonly transportation. People, cars, and other vehicles can be transported along with a road network; commercial goods can be transported along with a logistic network; phone calls can be transported along with a telephone network, and water pollution can be transported along with a stream or river network. There are two types of network analysis functions: optimal pathfinding and network partitioning. Optimal path finding generates the least cost path on a network between two predefined locations using both geometric and attributes data. Network partitioning assigns network elements (nodes or line segments) to different locations based on predefined criteria.

9. GIS-based Multicriteria decision analysis in SWM

9.1 Overview of GIS-based MCDA

We have seen that the capability of GIS is reliant on spatial analysis functions such as overlay, connectivity, and proximity. These functions, however, always do not provide the best decision alternatives when there are complex and conflicting sets of criteria presents. For instance, in siting a new waste disposal site, the overlay function can be used to combine different factors such as proximity to a road, surface water, and groundwater, or site slope, elevation, and soil type. However, this function does not provide enough analytical support because of the limited capabilities for incorporating decision makers' preferences into the GIS-based decision-making process. Thus, the combination of geographic data and the decision maker's preferences into analysis for better output is required.

As a result, GIS can assist in minimizing value conflicts among conflicting interest parties by giving more and better information, whereas multicriteria decision-making analysis (MCDA) methodologies can aid in lowering factual disagreements [2]. MCDA is a strategy for assisting decision-makers through the essential process of establishing evaluation criteria and determining relevant values in a choice circumstance. Based on literature MCDA has six components as a primary goal, the decision-maker(s), set of criteria, decision alternatives, decision environment, and outcome. Furthermore, the components of MCDA can be achieved through the steps shown in **Figure 1**. As it is shown in the figure, any spatial decision problem can be structured into three major phases according to [45]; intelligence which examines the existence of a problem or the opportunity for change, design which determines the alternatives, and choice which decides the best alternative. The components and steps of MCDA were discussed in detail by [2, 46, 47].

9.2 Multi-Criteria Decision Aid Methods

According to a recent study aimed to present a literature review of MCDA applications used in SWM, the top five and most commonly used MCDA methods are Analytic Hierarchy Process (AHP), Simple Additive Weighting (SAW), Elimination and choice expressing the reality (ELECTRE), and Preference Ranking Organization Method for Enrichment (PROMETHEE) and Technique for Order of Preference by Similarity to Ideal Solution (TOPSIS). **Table 3** shows the commonly used MCDA

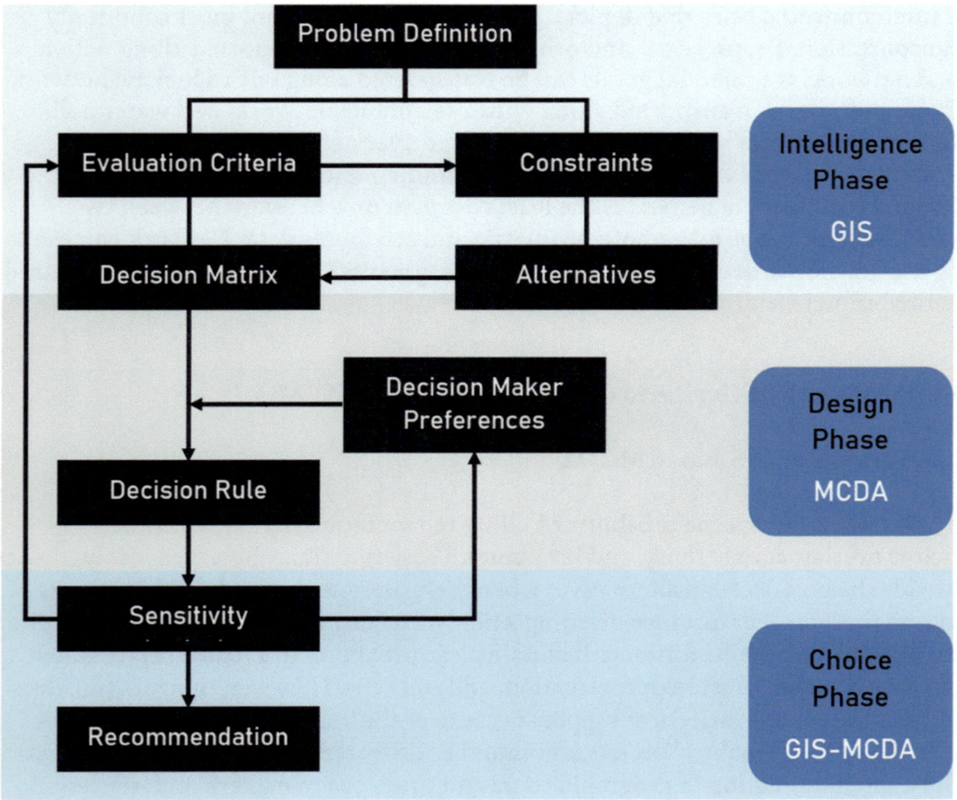

Figure 1.
Framework for spatial multicriteria decision analysis based on [2].

MCDA method	Description	Advantage	Disadvantage
SAW	Value based method Use of measurement of the utility of an alternative	Easy to use and well understandable. Applicable when exact and total information is collected. Well-proven technique. Good performance when compared with more sophisticated methods.	Normalization is required to solve multidimensional problems
AHP	Use of value-based, compensatory, and pairwise comparison approaches. Use of Hierarchical structure to present complex decision problem	Applicable when exact and total information is collected. A decision problem can be fragmented into its smallest elements, making evidence of each criterion applied. Applicable for either single or multiple problems, since it incorporates qualitative and quantitative criteria. Generation of inconsistency index to assure decision-makers	Due to aggregation, compensation between good scores on some criteria and bad scores on other criteria can occur Implementation is quite inconvenient due to the complexity Complex computation is required Time-consuming
TOPSIS	Use of the value-based compensatory method Measures the distances of the alternatives from the ideal solution Selection of the one closest to the ideal solution	Easy to implement the understandable principle Applicable when exact and total information is collected Consideration of both the positive and negative ideal solutions Provision of a well-structured analytical framework for alternatives ranking Use of fuzzy numbers to deal with uncertainty problems	• Normalization is required to solve multidimensional problems
ELECTRE	Use of outranking method Use of pairwise comparison, compensatory Use of indirect method that ranks alternatives utilizing pairwise comparison	Applicable even when there is missing information Applicable even when there are incomparable alternatives Applicable even when the incorporation of uncertainties is required	Time-consuming without using specific software due to complex computational procedure May or may not reach the preferred alternative

MCDA method	Description	Advantage	Disadvantage
PROMETHEE	Use of outranking method, pairwise comparison, and compensatory method Use of positive and negative preference flows for each alternative in the valued outranking Applicable even when simple and efficient information is needed Generation of ranking with decision weights	Applicable even when there is missing information	Time-consuming without using specific software When using many criteria, it becomes difficult for decision-makers to obtain a clear view of the problem

Table 3.
Comparison of MCDA methodologies applied to SWM based on [48].

methodologies in SWM, their description, advantages, and disadvantages of these methodologies.

As mentioned in **Table 3**, we have to understand that each MCDA available has its strength and drawback, and also there is no general rule in adopting one. Therefore, in SWM the decision-maker decides to depend on the criteria and alternatives available. This also depends on the decision-maker's previous experience and the availability of adequate software [47].

9.3 Criteria for SWM planning

As discussed in previous sections, GIS-based MCDA is used to overcome complex and conflicting criteria in SWM. In fact, in MCDA methodology development there are goals and alternatives where decision outputs are made based on the decision-maker preferences. In addition, in sustainable waste management or ISWM, the aim is to reduce waste generation, improve reuse, recycle, and recovery, and otherwise properly manage these wastes. So, without any compromises, there are always environmental, economic, political, and social considerations in ISWM. Due to these facts, ISWM planning has been a challenge for the decision-makers as different criteria are taken into consideration under the specific goal of sustainable solid waste management. For instance, [49] used sixteen criteria for landfill site selection, including topography, urban and rural settlements, highways and village roads, railways, airports, wetlands, pipelines, and power line infrastructure, slope, geology, land use, floodplains, aquifers, and surface water. In general, the criteria used in ISWM planning can be categorized as environmental criteria, political criteria, financial and economic criteria, hydrologic and hydrogeologic criteria, topographical criteria, geological criteria, availability of construction materials, and other criteria as shown in **Figure 2**.

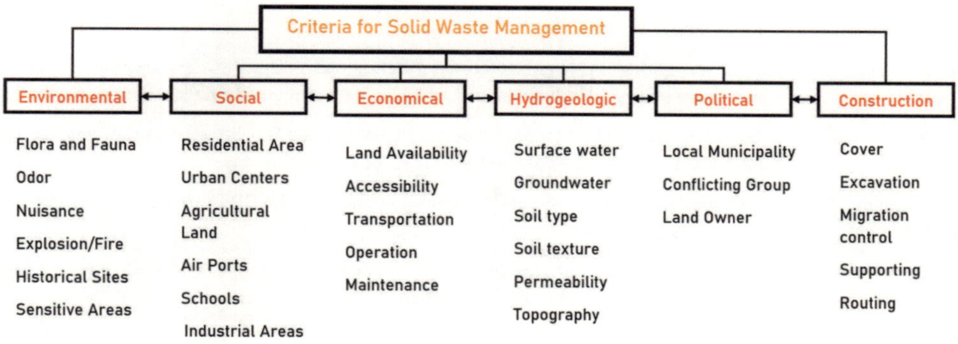

Figure 2.
Criteria used for decision making in SWM.

10. Application of GIS in optimizing solid waste collection

10.1 Routing and waste transportation

Routing in solid waste collection entails planning and specifying routes for trucks to follow during the collection process [50]. When it comes to SWM optimization, routing is crucial. The rate of solid waste collection is frequently determined by the efficiency of transportation in the SWM component. Due to their complicated structures, transportation of solid waste to the final disposal or treatment facility is a major problem in many cities. As a result, more emphasis should be placed on optimizing waste collection routes, as failure to do so would result in exorbitant costs. As a result, defining the routes reduces waste collection costs while delivering the best service to the community. Routing is essentially the process of choosing a path for traffic within a network, as well as between or across various networks. The five basic steps of route planning are as follows, and they are accomplished in GIS utilizing the previously mentioned principles: (a) Identifying the potential location, (b) Identifying the storage capacity and volume to be collected, (c) Grouping the potential locations for a single truck cover, (d) Planning the shortest route between different groups, (e) Choosing the optimum route that is a shorter distance, less traffic volume, and less expensive.

11. Strategies for route optimization

Over time, routing has been applied in varying ways in solid waste management, giving rise to many models and strategies. According to EPA categorization, these models can be divided into macro-routing, districting and route balancing, and micro-routing [51]. The major difference is that macro-routing aims to optimize the use of the waste collection in daily and long-term capacity by minimizing round trip and haul time. While districting and route balancing divide the workload between the workers, micro-routing considers the details of each daily waste collection so that unnecessary truck movements will be reduced [51]. Among the many identified strategies for route optimization, heuristic routing, right turns, onboard computers, and round trips were presented in **Table 4**.

Strategies	Description
Heuristic Routing	Heuristic routing is a system used to describe how deliveries are made when problems in a network topology arise. In heuristic routing, routes should not be fragmented or overlapping
Right Turns	Using only right turn during the collection of solid waste. This helped not only reduce cost by reducing fuel consumption but promoted the safety of drivers.
Computer-based	This allows the collectors to track routes in realtime. It is also GPS-based and every detail of the route is presented.
Dump trips	Scheduling the collection trips for the full containers twice per week. This also involves when the containers are full, they should be taken by the nearby truck.

Table 4.
Strategies to optimize the routing in the solid waste management.

12. High-Density Routing and Point-to-Point Routing

Residential collections are often routed using arc routing or side-of-street routing in high-density routing. This enables the software to arrange automatic collection with two passes on a street segment, as well as a semi-automated collection to serve both sides of a street segment at the same time. Point-to-point routing can be done on a variety of software and web-based platforms. When using point-to-point, the collected side of the roadway is usually ignored. Point-to-point routing can be used to develop commercial routes for solid waste collection as well as for routing calls for services like cart or bin delivery.

13. Waste bin location

For waste to be collected efficiently, proper waste bin allocation and distribution are critical. It simplifies waste sorting, recycling, and transportation at the source. As a result, the second functional element of solid waste management is waste storage at the source. The waste is usually placed in bins on both sides of streets, close to buildings and other sources of waste generation. As a result, due to health concerns, attention should be exercised when storing hazardous material near residential areas. Wastes should also be stripped away regularly. GIS can be used to appropriately place waste storage containers or waste bins so that they are no longer a menace, are evenly distributed among the households, and the cost is decreased.

Several criteria were used to determine where the waste bins should be placed, including proximity to a road and a waste-producing source, land usage, sensitive areas, and so on. Using GIS, the criteria are integrated with the preferences of the decision-makers to obtain optimized waste bin sites using spatial analysis such as buffering. For example, a decision-maker could choose a buffer zone that is 20 meters away from a waste-producing source, 10 meters away from roadways, but 100 meters away from sensitive places such as hospitals, historical sites, and schools. The number of waste bins allocated can also be decided based on the waste generation rate and dispersed evenly. Furthermore, one of the most common strategies for arranging waste bins is using Location-Allocation models to determine the best position. The p-median model is used in several investigations. Because it averages the locations of multiple points, this model decreases the distance and expense of the facility from the source.

14. Application of GIS in waste disposal planning

In waste disposal planning, a geographic information system plays an important role. GIS is used in waste disposal for a variety of reasons, including decision support for locating suitable landfills and temporal monitoring of disposal locations, including landfills. Due to the competing criteria illustrated in **Figure 2**, solid waste landfill siting is a complex and time-consuming operation in the traditional method of SWM. Furthermore, the primary purpose of the landfill site selection procedure is to ensure that the disposal facility is located in the best possible area, with the least amount of detrimental influence on the environment and population. Furthermore, a thorough review process is required to determine the best possible disposal place that complies with government standards while also minimizing costs.

Typically, landfilling is the least preferred method of ISWM according to the waste management hierarchy. But proper management of residue from reuse, recycling, and recovery, as well as the ashes from incineration, is a must. Therefore, the GIS-based MCDA approach is common in landfill site selection analysis. The basics and principles of GIS-based MCDA have been discussed in previous sections, so in this section, an overview of how GIS can be applied to landfill suitability analysis is presented. **Figure 3** depicts the methodology for adopting GIS, MCDA, and waste

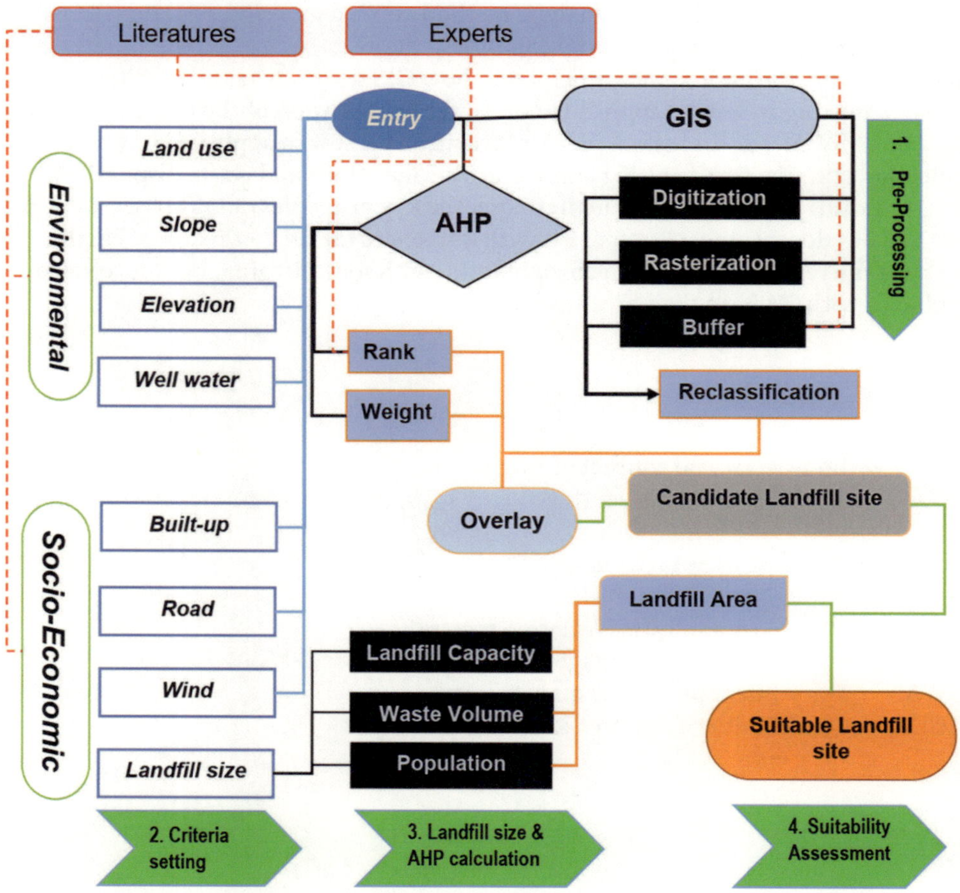

Figure 3.
Framework of the study to select a suitable sanitary landfill site, adopted from [41].

disposal site analysis. Also, as presented in **Figure 2**, the process always starts with setting a goal (suitable landfill site). Then, after defining a set of criteria, GIS functions are combined with decision makers' preferences to produce the final, most suitable landfill site.

15. Conclusion

This chapter deals with the application of geographic information systems to solid waste management. GIS is a powerful tool that can assist in minimizing value conflicts among conflicting interest parties by giving more and better information. Essentially, planning a sustainable waste management approach is complex, tedious, and time-consuming, and decision-makers are frequently subjected to conflicting factors in SWM planning. There is an increasing trend of waste generation worldwide, and the situation is worse in developing countries owing to poor infrastructure, finance, and political reasons. To cope with the high waste generation and different problems in SWM, integrated solid waste management is widely used as a sustainable waste management practice. ISWM is a complex and tedious process to implement. Following the challenges in ISWM, many researchers came up with many innovations, like mathematical modeling and computer-based modeling, to mention a few. GIS is a computer-based spatial analysis method applied to SWM, enabling decision-makers to make better judgments by combining the alternatives and their preferences. In this chapter, the basics, and principles of how GIS works, what multicriteria decision making is, how to apply it to GIS, and the utilization of the GIS-based MCDA method in SWM was discussed. Also, information on how to apply GIS to waste collection optimization (routing and waste bin allocation) and waste disposal planning (landfill) was presented. Furthermore, the key principles of GIS-based MCDA method development were supported with references for further reading. Finally, we hope the readers will get some insights and, with some digging, be able to solve problems with SWM in their area easily.

Conflict of interest

The authors declare no conflict of interest.

Acronyms and abbreviations

GIS	Geographic information system
SWM	Solid waste management
ISWM	Integrated waste management
MCDA	Multicriteria decision making analysis

Author details

Elsai Mati Asefa*, Kefelegn Bayu Barasa and Dechasa Adare Mengistu
Department of Environmental Health, College of Health and Medical Science, Haramaya University, Harar, Ethiopia

*Address all correspondence to: elsyyymati@gmail.com

References

[1] UNEP. Guidelines for National Waste Management Strategies: Moving from challenges to opportunities. Nairobi: United Nations Environment Programme. UNEP, Nairobi; 2013

[2] Leake C, Malczewski J. GIS and Multicriteria Decision Analysis. Vol. 51. Hoboken, New Jersey, U.S.: John Wiley & Sons; 1999. p. 247

[3] USEPA. Criteria for the Definition of Solid Waste and Solid and Hazardous Waste Exclusions. In: What is a Solid Waste? US EPA; 2021. pp. 1-12. Available from: https://www.epa.gov/hw/criteria-definition-solid-waste-and-solid-and-hazardous-waste-exclusions

[4] Hoornweg D, Thomas L. What a waste: solid waste management in Asia. The World Bank; 1999

[5] Tchobanoglous G. Solid waste management. In: Environmental Engineering Environmental Health Safety Municipal infrastructure L use planning. New Jersey: Ind Wiley; 2009. pp. 177-307

[6] Tchobanoglous G, Frank K. Handbook of solid waste management. New York: The McGraw-Hill Companies, Inc; 2002

[7] US EPA. What Is Integrated Solid Waste Management? Vol. 6. Washington DC: United States Environmental Protection Agency; 2002. pp. 1-4. Available from: www.epa.gov/globalwarming

[8] McDougall FR, White PR, Franke M, Hindle P. Integrated solid waste management: A life cycle inventory: Second edition. New Jersey: John Wiley & Sons, Inc; 2008. pp. 1-513

[9] ZWIA. How Communities Have Defined Zero Waste | US EPA [Internet]. Washington DC: Zero Waste International Alliance; 2018. Available from: https://www.epa.gov/transforming-waste-tool/how-communities-have-defined-zero-waste

[10] Conserve Energy Future. Sustainable Practices in Waste Management [Internet]. New York: Conserve Energy Future. 2021. Available from: https://www.conserve-energy-future.com/sustainable-practices-waste-management.php#:~:text=Sustainable waste management is a, source of energy and resources

[11] Hussain CM. Concepts of Advanced Zero Waste Tools Present and Emerging Waste Management Practices. Moscow, US: Electronic Green Journal; 2020

[12] Munguía-López A d C, Zavala VM, Santibañez-Aguilar JE, Ponce-Ortega JM. Optimization of municipal solid waste management using a coordinated framework. Waste Management. 2020;**115**:15-24

[13] de Souza V, Melaré A, Montenegro González S, Faceli K, Casadei V. Technologies and decision support systems to aid solid-waste management: a systematic review. Waste Management. 2017;**59**:567-584

[14] Kundariya N, Mohanty SS, Varjani S, Hao Ngo HWC, Wong J, Taherzadeh MJ, et al. A review on integrated approaches for municipal solid waste for environmental and economical relevance: Monitoring tools, technologies, and strategic innovations. Bioresource Technology. 2021;**342**:125982

[15] Sibanda LK, Obange N, Awuor FO. Challenges of Solid Waste Management

in Kisumu, Kenya. Urban Forum. Switzerland: Springer; 2017. pp. 387-402

[16] El-Fadel M, Findikakis AN, Leckie JO. Environmental impacts of solid waste landfilling. Journal of Environmental Management. 1997;**50**(1):1-25

[17] Wang Z, Lv J, Gu F, Yang J, Guo J. Environmental and economic performance of an integrated municipal solid waste treatment: A Chinese case study. Science of Total Environment. 2020;**709**:136096. Available from: https://www.sciencedirect.com/science/article/pii/S0048969719360929

[18] Gaeta GL, Ghinoi S, Silvestri F, Tassinari M. Innovation in the solid waste management industry: Integrating neoclassical and complexity theory perspectives. Waste Management. 2021;**120**:50-58. Available from: https://www.sciencedirect.com/science/article/pii/S0956053X20306255

[19] Hafid HS, Omar FN, Abdul Rahman N, Wakisaka M. Innovative conversion of food waste into biofuel in integrated waste management system. Critical Reviews in Environmental Science and Technology. 2021;**51**:1-40

[20] Colangelo F, Farina I, Travaglioni M, Salzano C, Cioffi R, Petrillo A. Eco-efficient industrial waste recycling for the manufacturing of fibre reinforced innovative geopolymer mortars: Integrated waste management and green product development through LCA. Journal of Cleaner Production. 2021;**312**:127777

[21] Paul K, Chattopadhyay S, Dutta A, Krishna AP, Ray S. A comprehensive optimization model for integrated solid waste management system: A case study. Environmental Engineering Research. 2019;**24**(2):220-237

[22] Anderson LE, Nigam A. A mathematical model for the optimization of a waste management system. Vol. 1. Berkeley, USA: University of California, Sanitary Engineering Research Laboratory SERL Rep.; 1968. p. 68

[23] Fiorucci P, Minciardi R, Robba M, Sacile R. Solid waste management in urban areas: Development and application of a decision support system. Resources, Conservation and Recycling. 2003;**37**(4):301-328

[24] Costi P, Minciardi R, Robba M, Rovatti M, Sacile R. An environmentally sustainable decision model for urban solid waste management. Waste Management. 2004;**24**(3):277-295

[25] Abou Najm M, El-Fadel M. Computer-based interface for an integrated solid waste management optimization model. Environmental Modelling & Software. 2004;**19**(12):1151-1164

[26] Singh A. Remote sensing and GIS applications for municipal waste management. Journal of Environmental Management. 2019;**243**:22-29

[27] Waters N. GIS: History. In: International Encyclopedia of Geography. Vol. 15 Volume Set: People, the Earth, Environment and Technology. New Jersey: John Wiley & Sons, Inc.; 2018. pp. 1-13

[28] Coppock TJ, Rhind WD. The history of GIS. Geographic Information System Vol 1 Princ. 1993;21-43.

[29] McHarg IL. Design With Nature. New York: American Museum of Natural History; 1969. pp. 7-17

[30] By RA d, Huisman O. Principles of geographic information systems. In: TS-WorldCat T4-An introductory

textbook. Vol. 1. Atlanta, Georgia: ITC Educ Textb Ser; 2009. p. 258. Available from: http://www.worldcat.org/oclc/700395896

[31] López AJV, Aguilar Larrucea M, Fernández-Carrión Quero S, Jiménez del Valle A. Optimizing the collection of used paper from small businesses through GIS techniques: The Leganés case (Madrid, Spain). Waste Management. 2008;**28**(2):282-293

[32] Chang N-B, Lu HY, Wei YL. GIS technology for vehicle routing and scheduling in solid waste collection systems. Journal of Environmental Engineering. 1997;**123**(9):901-910

[33] Hina SM, Szmerekovsky J, Lee ES, Amin M, Arooj S. Effective municipal solid waste collection using geospatial information systems for transportation: A case study of two metropolitan cities in Pakistan. Research in Transportation Economics. 2020;**84**:100950

[34] Dao-Tuan A, Nguyen-Thi-Ngoc A, Nguyen-Trong K, Bui-Tuan A, Dinh-Thi-Hai V. Optimizing vehicle routing with path and carbon dioxide emission for municipal solid waste collection in Ha Giang Vietnam. Lect Notes Inst Comput Sci Soc Telecommun Eng LNICST. 2018;**221**:212-227

[35] Vu HL, Ng KTW, Bolingbroke D. Parameter interrelationships in a dual phase GIS-based municipal solid waste collection model. Waste Management. 2018;**78**:258-270

[36] Van Mai C, Nguyen SH, Dao CD, Van LT, Phan HA. A GIS application in optimizing the collection and transportation route of domestic solid waste in Hue city Vietnam. In: Global Changes and Sustainable Development in Asian Emerging Market Economies. Vol. 2. Cham: Springer; 2022. pp. 599-609

[37] Chaudhary S, Nidhi C, Rawal NR. Gis-based model for optimal collection and transportation system for solid waste in allahabad city. Advances in Intelligent Systems and Computing. 2019;**814**:45-65

[38] Chang LYT. Optimal siting of transfer station locations in a metropolitan solid waste management system. Spectroscopy Letters. 1996;**30**(3):601-623

[39] El-Hallaq MA, Mosabeh R. Optimization of municipal solid waste management of bins using GIS. A Case Study: Nuseirat City. Journal of Geographic Information System. 2019;**11**(01):32-43

[40] Muttiah RS, Engel BA, Jones DD. Waste disposal site selection using gis-based simulated annealing. Computational Geosciences. 1996;**22**(9):1013-1017

[41] Asefa EM, Damtew YT, Barasa KB. Landfill site selection using GIS based multicriteria evaluation technique in Harar City, Eastern Ethiopia. Environ Health Insights. 2021;**15**:117

[42] Rahimi S, Hafezalkotob A, Monavari SM, Hafezalkotob A, Rahimi R. Sustainable landfill site selection for municipal solid waste based on a hybrid decision-making approach: Fuzzy group BWM-MULTIMOORA-GIS. Journal of Cleaner Production. 2020;**248**:119186

[43] Rolf A, de By RA. Principles of geographic information systems. Vol. 99. Netherland: Int: Inst Aerosp Surv Earth Sci (ITC) Hengelosestraat; 2001

[44] Aronoff S. Geographic information systems: A management perspective. In: Geographic information systems: A management perspective. Vol. 4. 1989. p. 58

[45] Simon HA. The new science of management decisions. New York: Harper & Row; 1960

[46] Malczewski J, Rinner C. Multicriteria decision analysis in geographic information science. Vol. 1. Switzerland: Springer; 2015

[47] Pires A, Martinho G, Rodrigues S, Gomes MI. Sustainable solid waste collection and management. Switzerland: Springer; 2019

[48] Pires A, Bin CN, Martinho G. An AHP-based fuzzy interval TOPSIS assessment for sustainable expansion of the solid waste management system in Setúbal Peninsula, Portugal. Resources, Conservation and Recycling. 2011;**56**(1):7-21

[49] Sener B, Süzen ML, Doyuran V. Landfill site selection by using geographic information systems. Environmental Geology. 2006;**49**(3): 376-388

[50] Sulemana A, Donkor EA, Forkuo EK, Oduro-Kwarteng S. Optimal Routing of Solid Waste Collection Trucks A Review of Methods. Journal of Engineering. 2018;**2018**:56

[51] Shuster KA, Schur DA. Heuristic routing for solid waste collection vehicles. Vol. 45. Washington DC: USA EPA; 1974

Section 6

Impact of COVID-19 Measures on the Air Quality

Chapter 8

Impact of COVID-19 Measures on the Air Quality Monitored for the State of Himachal Pradesh: A Google Earth Engine Based Study

Abhinav Galodha, Chander Prakash and Devansh Raniwala

Abstract

The COVID-19 pandemic was declared by World Health Organization (WHO) on 11 March 2020 and advised countries to take immediate and concerted action. The governments of India and Himachal Pradesh carried out preventive and precautionary steps to minimize the spread of coronavirus disease. In this study, the impact of a sudden halt in human activity on air quality was investigated by looking at changes in satellite imagery using a remote sensing approach. The concentrations of the gaseous contaminants studied (CO, SO_2, NO_2, and C_6H_6) show a significant decrease during the lockdown. The average particulate matter concentrations (PM_{10} and $PM_{2.5}$) differed significantly from gaseous emissions, meaning that particulate matter significantly affects anthropogenic activities. NO_2 concentrations and NOx emission variations were tracked for rural/town areas around Himachal Pradesh and major urban cities of India. Daily top-down NOx emissions were measured using the Tropospheric Monitoring Instrument (TROPOMI), which assisted in retrieving NO_2 from the steady-state continuity equation. The emissions of NO_x from rural, urban, and power plants were compared before and after the lockdown. The research accounted for our studies on the levels of (NO_2, Ozone (O_3), and sulfur dioxide (SO_2) were monitored using Sentinel-5P imagery using the GEE platform.

Keywords: Google earth engine (GEE), TROPOMI, sentinel -5P, COVID-19, WHO, SDG, NO_2, O_3, SO_2

1. Introduction

Dissolved gases in the air, especially (O_2, O_3, N_2, and so on) are valuable and essential resources that help sustain life on earth. There is ever-increasing stress on openly accessible air resources due to the atmospheric pollution caused due to the advent of the use of fossil fuels, industrial discharges, and transport traffic, with only one factor responsible for it: the rise in population. An ever-increasing pressure mounting on them due to heavy dependence on fossil fuels leads to an unprecedented health crisis ranging from local to national levels [1, 2]. Air is vital to sustain and flourish the planet's health and the environment [3, 4]. However, with rapid globalization, urbanization and industrialization, there has been an enormous crisis with unhealthy air quality levels leading to vulnerable ecosystems [5]. Poor air quality in developed Western countries leads to roughly 60,000 deaths annually. The economic costs and repercussions are massive staggerings at $150 – $ 200 billion in prices leading to high levels of air pollutants. Equally important information that supports life in an ecosystem is delivered by ambient air quality resources [6]. An increase in air pollution hampers and deteriorates ambient air quality and threatens human health, aerial ecosystem balance, economic development, and social well-being [7, 8].

According to the World Health Organization (WHO), 90% of the world population lives in harsh to dangerously polluted places, breathing high levels of air containing highly high levels/rates of pollutant concentrations, and 5–7 million deaths occur every year as a result of exposure to ambient air pollution and from the exposure to smoke from fuels such as wood and fossil fuels [5, 8]. As described earlier, the population's health degradation occurs due to the onset and remission of harmful industrial pollutants. Nitrogen dioxide is one such greenhouse gas (GHG) that is reddish and results from NO conversion in the presence of volatile organic compounds (OVCs) [9]. NO_2 monitoring is quintessential because of the potential threat they hold:

1. NO_2 being a primary pollutant, is a significant cause of the creation of secondary pollutants such as peroxyacetyl nitrates (PAN), ozone (O_3), and nitric acid (HNO_3).

2. Visibility reduction in urban areas.

3. Negative impacts on human health.

The onset of the efforts that were carried out to limit the transmission of the SARS-CoV-2 virus that led to an unprecedented havoc and pandemic situation across the globe for roughly two years was minimized by carrying out strict lockdown implementations. These policies, in hindsight, had a more considerable extent impacting the day-to-day activities which confined the public within their homes, reducing human activities, especially in the industrial and transport sectors [10]. This led to a significant decrease in air pollution concentration levels. Like NO_2, other greenhouse gases (GHGs) have a dampening and profound negative effect, impacting the overall health cycle of ambient air qualitative concentrations [4, 7].

Air quality resource management requires a continuous and accurate monitoring assessment to support a real-time network. Satellite remote sensing observations [1, 9] have provided real-time and continuous data that have been beneficial in tracking across several years [11] and have served in a time- and cost-effective manner for carrying out large-scale monitoring [2, 3]. Air pollution is an important environmental issue, which needs to be addressed with solid policy implementations with technical guidelines and administrative protocols. For measurements across space and time, remote sensing satellite observation can play a pivotal role in atmospheric measurements related to air quality. As a result of the European Space Agency's close collaboration (ESA) with the Netherlands, the European Commission, EU industry, data users, and scientists came together for the Copernicus Sentinel-5 Precursor mission (Sentinel-5P). Sentinel-5P satellite was successfully launched from the Russian Federation (Plesetsk cosmodrome) on 13 October 2017. The main objective of the Copernicus Sentinel-5P's mission is to perform atmospheric measurements with a high spatio-temporal resolution for air quality, ozone & UV radiation, and climate monitoring & forecasting. The task consists of one satellite carrying the Tropospheric Monitoring Instrument (TROPOMI). The TROPOMI instrument was co-funded by ESA and The Netherlands. The TROPOMI [10, 12] was on board with the 49 European Copernicus Sentinel-5 Precursor (S5P) satellite designed explicitly for 50 tropospheric monitoring on the global scale and has a daily revisit time. If we compare this to its predecessor OMI, TROPOMI's spatial resolution (3.5 x 5.5 km^2) is roughly 15 times better, and its signal-to-noise ratio (SNR ratio) per ground pixel is much more dominant. For relevant air quality products, including NO_2, SO_2, HCHO, and CHOCHO, this results in a spectacular improvement in measurement sensitivity, thus enabling the study of rapid emission changes for even smaller sources than existing options. The daily global coverage of TROPOMI for CO measurements is at a resolution of 7 x 5.5 km^2, representing a massive improvement from its predecessor SCIAMACHY [13], especially with its spatial resolution.

The absorbent dissolved gases in the air and the estimations from Sentinel-5P provide an opportunity to observe the magnitude and timing of the changes in tropospheric trace gas constituents resulting from unprecedented COVID-19 lockdown measures. The initial TROPOMI observations offer a template that looks into the dramatic reduction and significant changes in the NO_2 concentrations over regions with strictly enforced multiple lockdowns and mini-cluster zones in important cities for pan India. This triggered a high level of interest across the globe and initiated a massive scale of approximately 60 plus studies, initially starting at a global scale and then mainly aimed at a regional scale and finally, primarily focused on GHG emissions estimation [4]. However, the unparalleled capacity of TROPOMI to provide relevant information on COVID-19-driven emission reductions based on multiple measurements has not been exploited at a regional scale as far as was expected. The present research is to estimate the COVID-19-driven changes in the concentration of a few major trace gases (CHOCHO, NO_2, SO_2, CO, HCHO, etc.) from the regional scale for a region in the North of India, Himachal Pradesh, surrounded by the Himalayas from the North and bordered by China in the east with the impact of lockdown restrictions and using state-of-the-art TROPOMI operational and scientific data products. In doing so, we further try to expand on the view that the unique capabilities provided by the TROPOMI instrument are helpful inconsistently tracking the changes in ambient air quality and anthropogenic emissions across the region of interest [7].

These gases have significant human-induced anthropogenic effects with their relative contribution to energy variations, industrial standards, and transport sector emissions [6]. Each sector responded at a different scale to the COVID-19 lockdown. Several TROPOMI trace gas products contain additional metadata on emissions emitted at scale level and the COVID-19 lockdown-induced impacts on the atmospheric structure. We show that meaningful visualization graphical plots, trends, and source plots can be obtained using the high spatial resolution of TROPOMI data. By taking the median composite of the study area. Although this is primarily the result of the improved response or sensitivity of the instrument, we also introduce new engagements and developments in trace gas retrieval techniques and tweaking to enhance the sensor sensitivity of the TROPOMI datasets to even smaller trace determinations and more minor changes in emissions. To achieve these goals, we discuss the way forward and the limitations of each of the estimations or retrievals for monitoring regional changes [9].

In the present section, we have described TROPOMI data in general terms, followed by the study area, the methodology used to address the retrieval process, and a detailed discussion and analysis of how we handle each data result in this study [8].

The goal of this methods and data section is not only to explain how this study was conducted but also to provide guidance to remote sensing analysts, GIS enthusiasts, and research domain interest holders on how to best interpret, understand and analyze TROPOMI trace gas data not only for lockdown-driven emission changes but also for other event-driven changes. The below section describes the study area and the impacts of COVID-19 lockdown measures on all continents, using TROPOMI NO_2 data. The following two areas will tell the effect of the lockdown measures on a regional scale by examining NO_2, SO_2, CO, HCHO, and CHOCHO for the North part of India. The last part of the book chapter will cater to the progressive outlook of future possibilities, the challenges, limitations, and a way forward for this type of analysis, followed by a conclusion and references [14].

2. Study area and methods

The study area chosen lies in the northern part of India in the North belt of the Himalayas in Himachal Pradesh. Himachal Pradesh is a province of snow-laden mountains that constitute India's north part. Situated in the Western Himalayas, it is one of the 13 mountain states and is characterized by an extreme landscape featuring several peaks and extensive river systems. Himachal Pradesh is the northernmost state of India and shares borders with the union territories of Jammu and Kashmir and Ladakh to the North, and the conditions of Punjab to the west, Haryana to the southwest, Uttarakhand to the southeast, and a very narrow border with Uttar Pradesh to the south. Himachal is in the western Himalayas, situated between 30°22′N and 33°12′N latitude and 75°47′E′ and 79°04′E longitude. They cover an area of 55,673 square kilometers (21,495 sq. mi). The drainage system of Himachal is composed both of rivers and glaciers. Himalayan rivers crisscross the entire mountain chain. Himachal Pradesh provides water to both the Indus and Ganges basins. The drainage systems of the region are the Chandra Bhaga or the Chenab, the Ravi, the Beas, the Sutlej, and the Yamuna. These rivers are perennial and are fed by snow and rainfall. An extensive cover of natural vegetation protects them. Four Punjab rivers flow through the state, three originating here. The state of Himachal Pradesh is divided into 12 districts and three significant sub-divisions: Shimla, Kanga, and Mandi. The districts are divided into 73 subdivisions, 78 blocks, and 172 Tehsils. The predominantly mountainous region comprising the present-day Himachal Pradesh has been inhabited since prehistoric times, witnessing multiple waves of human migrations from other areas. Himachal Pradesh is spread across valleys with many perennial rivers flowing through them.

Roughly 90% of the state's population lives in rural areas. Agriculture, horticulture, hydropower, and tourism are essential sectors contributing to the state's economy. The hilly state is almost universally electrified, with 99.5% of the households having electricity as of 2016. The state was declared India's second open-defecation-free state in 2016. According to a survey of CMS – India Corruption Study 2017, Himachal Pradesh is India's least corrupt state.

The below figure gives a complete description of the study area, and the scope of interest, which is highlighted in orange color is depicted and shown on the map (**Figure 1**).

In this research study, our interpretation, analysis, and research are primarily based on TROPOMI data for regional understandings of the defined study area and compared with other major cities. The period of interest was initiated during

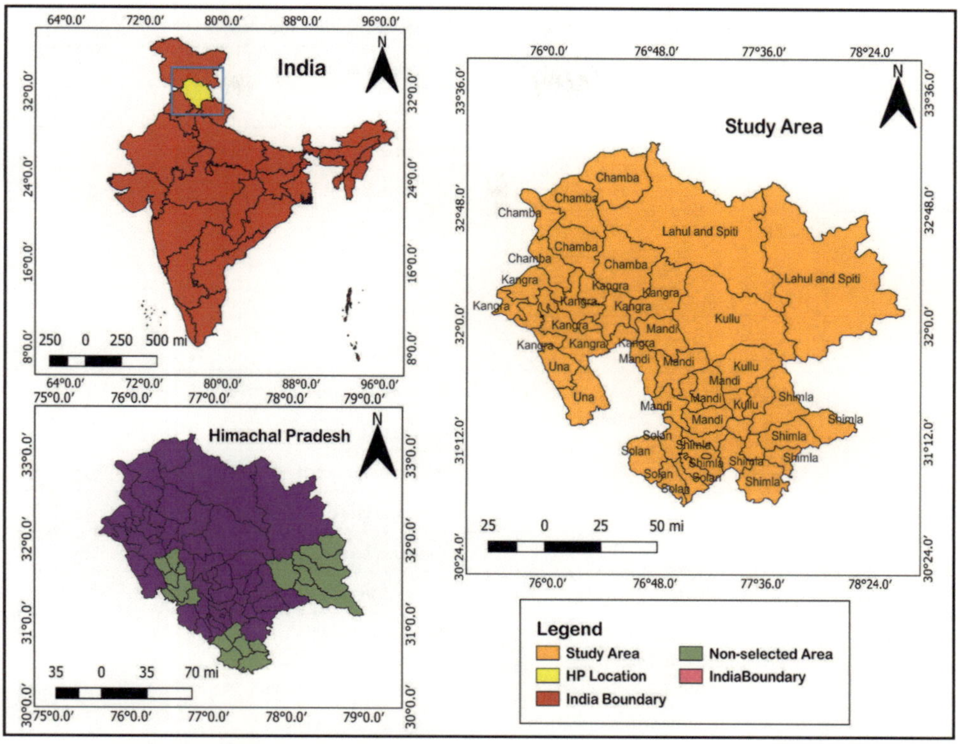

Figure 1.
Study area with the region highlighted on pan India map, Himachal Pradesh state location and boundary, the orange area shows a few of highlighted districts, for the part of interest (ROI), we have researched the entire state.

the lockdown starting from mid of March 2020 and then carried out a comparison concerning the pre-lockdown periods of 2019. The results and analysis are presented in the broader context of the TROPOMI operational data, which started in April 2018. The sensor we use for carrying out the observations is from the TROPOMI instrument, which was onboarded onto the S5P and categorized as a push-broom imaging spectrometer [10]. This helps to measure the ultraviolet (UV), visible (VIS), near-infrared (NIR), and shortwave infrared (SWIR) spectral band combinations, which were selected keeping in mind to cover the absorption features, i.e., water absorption features, cloud features absorption and with a presence of a large number of gaseous traces having atmospheric constituents. Using the spectral radiance measurements available from TROPOMI, atmospheric concentrations of gaseous trails are retrieved, and cloud and aerosol properties are determined. We use the following TROPOMI data products for this work: NO_2, SO_2, CO, HCHO, and CHOCHO, summarized in **Table 1**. We have tried to accommodate as many dissolved trace gases as we can. The S5P satellite flies in a sun-synchronous orbit, with a local overpass time of 13:30. TROPOMI has a 2600 km wide swath, providing near-daily global coverage. The spatial sampling of TROPOMI varies over the spectral bands [11]. The nadir/top-view sampling was at the start of the operational period in the initial period of 2018, which was approximately 3.5 x 7 km^2 (across- x along-track) for the ultraviolet (UV) and visible bands, and a 7 x 7 km^2 cross-section area covered by the shortwave infrared band (SWIR). In the second half of 2019, implementing a modified co-adding scheme, the sampling for these bands was thus, improved to 3.5 x 5.5 km^2 and 7 x 5.5 km^2, respectively [18].

Trace Gas	Spectral range	Lifetime	Primary Emission source
NO_2	405–465 nm	2 to 12 hours	• Transportation
[15]			• Industry
			• Power generation
			• Biomass burning
SO_2	310.25–326 nm	6 hours to several days	• Transportation
[16]			• Industry
			• Power generation
			• Volcanoes
CO	2324–2338 nm	Weeks to a month	• Transportation
[17]			• Industry
			• Power generation
			• Residential cooling and heating
			• Biomass burning
			• Oxidation of biogenic hydrocarbons
			• Methane oxidation
HCHO	328.5–359 nm	Several hours	• Primary and secondary product
[4]			• Biogenic emissions
			• Biomass burning
			• Industry
			• Transportation
CHOCHO	435–460 nm	2 to 3 hours	• Primary and secondary product
[12]			• Biogenic emissions
			• Biomass burning
			• Industry
			• Transportation

Table 1.
Summarizes the trace gases, their spectral range changes, atmospheric lifetime, and the primary emission sources for each trace gas addressed in this study.

Sentinel-5P observations are widely utilized within and beyond the remote sensing and GIS community. For the benefit of scientific users, it is crucial to provide information on how these observations can best be used, interpreted and analyzed [19]. The COVID-19 lockdown periods offer a unique use case for the Sentinel-5P lead model developers to highlight essential nuances in the individual atmospheric trace gases' lifetime and the detectability of each trace gas and show how these characteristics are critical to the interpretation of the highlighted observations. It is not sufficient, for example, to illustrate the lockdown-driven changes that affect the emissions simply by selecting a single day or week of TROPOMI column data for a given region as measured during a lockdown period for the same day or week from the year(s) prior [17]. We further address the impending challenges that need to be addressed and which are taken into cognizance, especially the description of the meteorological and seasonal variability from lockdown-driven pre and post-changes in trace gas emissions.

A brief tabulation and the sources of trace gas emissions and their lockdown-driven changes are depicted in an evaluative format. Generally, primary production trace gases, like NO_2 and SO_2, have a short life span and exhibit emission changes of utmost precision. Although NO_2 and SO_2 are primary producers of anthropogenic gaseous pollutants, the source sectors differ in each case [12]. For instance, the impact of lockdown on the power industry and the transportation sector was projected to significantly have a more considerable effect on NO_2 and SO_2 levels, as this sector is responsible for approximately 30% of the global NO_x emissions and about 1% of the global SO_2 emissions [4]. On the other hand, SO_2 emissions are most likely areas to be impacted by possible changes in power generation, as this sector accounts for 52% of the global SO_2 emissions and approx—30% of the worldwide NO_x emissions [4].

For CO, secondary production by methane oxidation and (biogenic) hydrocarbons accounts for at least 60% of the total atmospheric CO, followed by contributions from biomass burning and fossil fuel use [20]. Anthropogenic CO emissions originate from the industry, transportation, and residential sectors and account for about 30% of the global emissions [4]. Although local impacts of lockdown are likely for locations with anthropogenic solid CO emissions, overall, a much smaller lockdown-driven result is expected for CO based on its longer atmospheric lifetime and smaller contributions from lockdown-affected sources.

The trace gases of HCHO and CHOCHO are shortly living indicators of the non-methane volatile organic compound (NMVOC) categories resulting from an abiogenic process. The anthropogenic lead activity is significant to biomass exclusion and burning events [5]. They are primarily produced as secondary products from the oxidation of other NMVOCs but are also directly emitted from combustion and industrial processes, although to a lesser extent. In general, the relative production of CHOCHO from such combustion processes and the oxidation of aromatics, originating primarily from the industrial sector, is higher than for HCHO. Thus, the CHOCHO response to changes in anthropogenic emissions is expected to be more robust [17]. Retrievals provide information on the lower atmosphere/tropospheric level or total column that amounts to these gases because the spectra contain limited information on their vertical distribution in the atmosphere. TROPOMI observations thus provide a two-dimensional representation of the three-dimensional atmosphere [20]. The vertical profiles of each trace gas vary and significantly depend on the emissions' height and the trace gases' atmospheric lifetime (see **Table 1**). For example, NOx emissions at the surface result in NO_2 vertical profiles that peak in the near-surface layer (lowest 1–2 km of the troposphere) due to the short lifetime of NO_2. Similarly, SO_2 has a vertical profile that generally peaks in the lower troposphere.

On the other hand, CO has a lifetime period typically of weeks - to months and can be transported over great distances, both horizontally and vertically. Therefore, even though CO is often co-emitted with NO2, it has a significantly higher background concentration throughout the troposphere than NO_2. HCHO and CHOCHO have lifetimes of a few hours. Still, they are generally formed in the atmosphere via secondary production processes, leading to an intermediate profile shape compared to NO_2 and CO [21].

In addition to vertical profiles that vary per trace gas type, the vertical sensitivity of the TROPOMI measurements also varies across the variety and nature of trace gases. The sensitivity decreases towards the surface for the trace gases favorable and sensitive to the UV and VIS ranges. The accuracy of the retrieved column depends on a well-characterized a priori knowledge of the vertical distribution lead across it. Due to scattering, the near-surface sensitivity is lower in the UV (SO_2, HCHO) than in the VIS (NO_2 and CHOCHO) [22]. In the SWIR range, the vertical sensitivity is more constant and reaches well into the saturation stage. As part of the retrieval process, a priori vertical profiles of each trace gas are scaled to match the measured tropospheric column. Uncertainty in the retrieved column amount or sheer column density (VCD) is associated with inherent differences between the true and the available vertical profiles. However, the averaging kernels reported in the data products can be used to replace the a priori profiles with existing custom profiles [6], reducing the corresponding uncertainty. This study mainly focuses on changes in VCDs and uses standard a priori profiles for each data product. Therefore, the uncertainty related to the vertical profile is relatively small [13]. Another contribution to this error is partially cloudy scenes for each retrieval which raises the amount of unusable data and changes the vertical sensitivity. The cloud fraction threshold for each trace gas is described. In future studies, the averaging kernels could be used for inversion modeling existing emissions, thus, eliminating the limitations [23].

TROPOMI observes concentration changes of the emissions of the trace gases taken from the median or averaged out over a vertical column, which is not similar to the direct measurement of the near-surface emission. The column-averaged amount of a given trace gas measured at a specific location depends on emission, deposition, atmospheric changes, and photochemical reactions. Note that the background concentration is higher for trace gases with a longer atmospheric lifetime. In turn, enhanced background concentrations will increase the relative importance of atmospheric transport compared to local emissions. Local NO_2 emissions significantly impact At the same time, for CO, the contribution of remote sources can, in some cases, be superimposed on local emissions, thus making the interpretation more difficult. The effects of atmospheric climatic changes and chemical impact must also be encountered to attribute a change in concentration to a corresponding shift in local emissions [16].

3. The objective of the research and methodology

The major components of our research are collated into the following objectives, which are as listed below:

- To estimate the concentration changes in (CO, SO_2, NO_2, and Aerosol ($PM_{2.5}$, PM_{10}) from the Sentinel-5 Precursor mission assessing the air quality. The TROPOMI instrument is a multispectral sensor with a spatial resolution of 0.01 arc degrees.

- To compare the pollutant concentration changes from pre to post-lockdown conditions prevailing across the state.

- To visualize the trace gas concentration changes as part of geospatial air quality theme maps and do a trend analysis where gas concentration changes are mapped as a time duration function.

- To analyze the periodical changes in the absorbing aerosol index (AAI) across the one cycle period for the study area.

3.1 Methodology

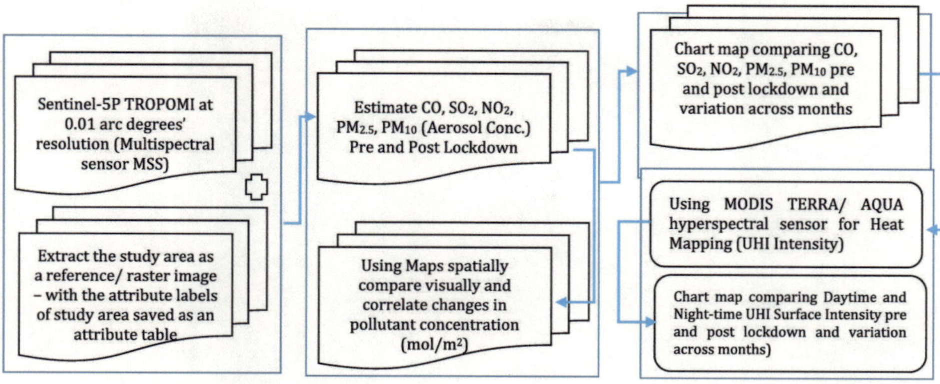

During the 1st phase of COVID-19, India carried out intensive and strict national lockdown measures limiting activities across the country, which had started on 24 March 2020, for 21 days to tackle the spread of the SARS-CoV-2 virus and protect its 1.3 billion inhabitants [24]. Careful region-based relaxations followed the stringent initial phase 1 restriction in three subsequent steps carried out through the end of May, as shown in **Table 2**:

Figure 2 gives an extent of TROPOMI observations of NO_2, SO_2, CO, HCHO, and CHOCHO, over India for April 2020, thus covering most phases 1 and 2 of the Indian lockdown compared to the same month in 2019. For NO_2 and SO_2, the concentrations are lower across the country in 2020 compared to 2019. Although less prominent, CO, HCHO, and CHOCHO appear to be lower in April 2020 over the Indo-Gangetic Plain (IGP) domain, which is one of the most densely populated areas of the world with roughly 900 million people [15].

NO_2 trace gas production sources are mainly from the traffic transport and energy-power sectors, roughly contributing about 30% of total anthropogenic emissions in India [4]. During phase 1 of the lockdown, the traffic had dropped by 80% [22], and energy consumption dropped by 25% compared to 2019 [3]. We expect a substantial reduction in NO_2, particularly in urban areas, due to significant decreases

Phase	Dates	Measures	Reference
Phase 1	24 March – 14 April	Nearly all services and factories were suspended.	[24]
Phase 2	15th April – 3 May	Extension of lockdown with relaxations, reopening of agricultural businesses and small shops at half capacity.	[5]
Phase 3	4th May – 17th May	The country is split into 3 zones: (i) lockdown zone, (ii) zone with movement with private and hired vehicles, and (iii) normal movement zone.	India today
Phase 4	17th May – 31st May	Additional relaxations, more authority are given to local bodies.	The Economic Times

Table 2.
Lockdown phases in India.

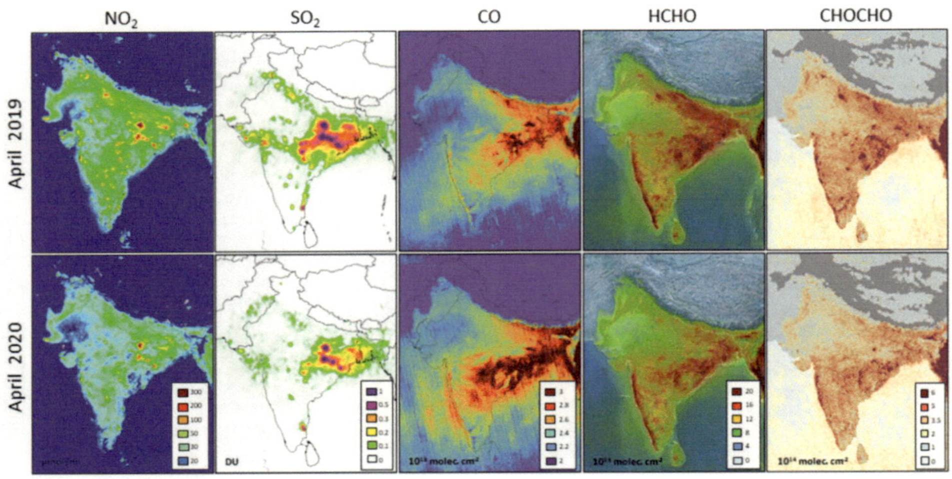

Figure 2.
The atmospheric trace gas variations between April 2019 and April 2020 for pan India. Image source: https://acp.copernicus.org/preprints/acp-2021-534/.

in transport sector activities. We also expect a weaker reduction near power plant corporations due to a smaller decrease in the energy demand. Indeed, as indicated by the maps of NO_2 column concentrations in **Figure 2**, a notable reduction in NO_2 can be seen in April 2020 compared to April 2019 [10]. An apparent reduction is observed in significant cities and the eastern part of India, where the most influential power plants are located. When both city centers and power plants are located within a 45 x 45 km^2 box, this box is excluded from the averages to avoid the potential outflow of one source to the other. A sharp reduction of 42% can be seen in the amount of NO_2 over cities during the first phase of the lockdown period starting at the end of March, compared to the same period in 2019. This initial drop in NO_2 concentration is followed by a gradual increase in concentration back to pre-lockdown levels with the successive relaxation phase implementations [25].

A sharp reduction of roughly ~45% in the amounts of NO_2 can be noticed for significant cities during the 1st phase of the lockdown period starting at the end of March, as compared to the same period in 2019. A slow but gradual increase follows this initial drop in NO2 in line with the successive relaxation phase. Power creation is a significant source of NO_2 in India, mainly observed in coal-fired power plants. When examining the average amount of NO_2 over the few largest coal-fired power plants, the observation was the significant decrease in NO2 during phase 1 of the lockdown period [26]. The fall in the observed coal-fired power plant sector is 23% compared to 2019 data, far less than the observed fall in NO_2 over large cities. Sentinel- 5P also observed an overall reduction in NO_2 with the shutting down of coal-power plants which were in line with the initial 25% decrease in the overload of electricity demand as reported by the National Load Dispatch Centre (NLDC) during phase 1 and lowering it to an 8% decrease during final steps of the lockdown (Fig. D1, [3]) (**Figure 3**).

According to the latest information from the emission inventory of 2019, the significant sources of SO_2 emissions in India are power generation (65%) and industry (25%) [4]. Since India largely depends on coal to fulfill its energy production, it is now the world's top emitter of anthropogenic SO_2 [19]. So, most of the SO_2 signal we observe in Sentinel- 5P data for this region is from coal-fired power plants, where

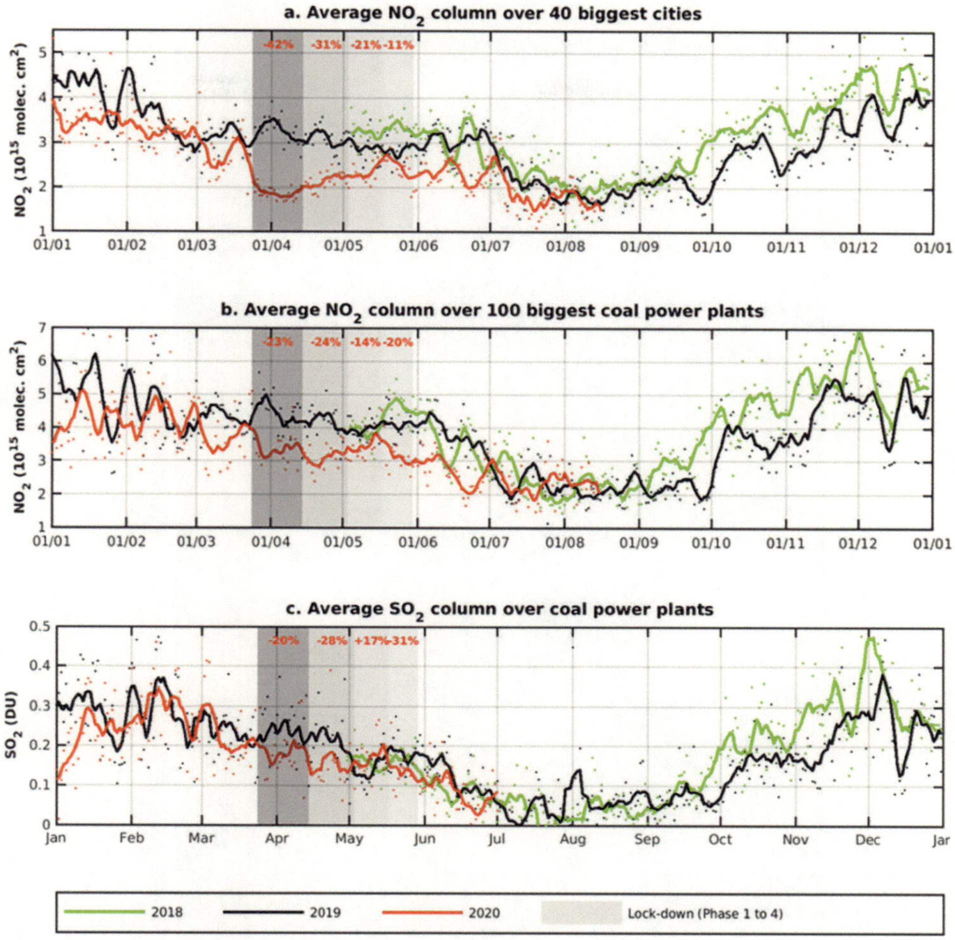

Figure 3.
Average tropospheric NO_2 concentrations for may 2018 (green), 2019 (black) until June 2020 (red) over the few largest Indian cities (top); and the few largest power plants in India (middle); and average SO_2 concentrations over the largest SO_2-emitting power plants in India. The different gray shading denotes the four different phases of the lockdown period. For each step, the reductions in NO_2 (or SO_2) concentrations are given relative to the definite periods in 2019. The dots are the daily means, and the solid lines represent the 7-day mean values [27]. Image source: https://acp.copernicus.org/preprints/acp-2021-534/.

contributions from conventional fossil fuels in India comprise a minor part of the response [3]. A reduction in SO_2 is visible over most areas. It is especially noticeable for the central-eastern portion of India, India's largest SO_2-emitting region, with much more than the coal-fired power plants available at other locations [28].

4. Results and discussion

The previous section dealt with changes in trace gas concentrations at the pan India scale. Still, this section will analyze the trace gas concentrations at the local and regional rankings for the study area region for the state of Himachal Pradesh. The work was carried out for the mentioned study area in the pre-lockdown and the

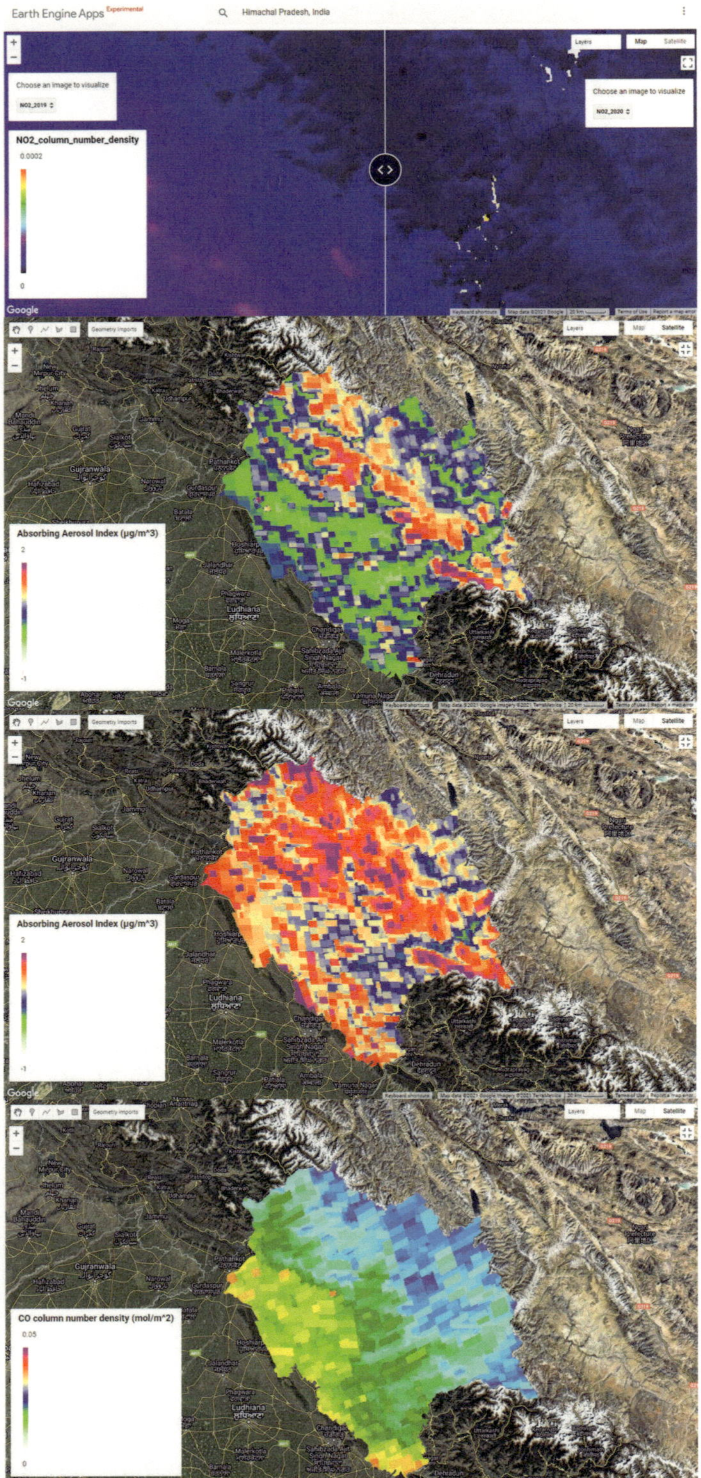

Figure 4.
The trace gas variability of CO (Mol/m2), NO_2 (Mol/m2), and absorbing aerosol index (AAI) (µg/m3) is depicted for both 2019 and 2020.

post-lockdown steps, divided into 4 phases. The trace gas variability of CO (mol/m2), NO_2 (mol/m2), and Absorbing Aerosol Index (AAI) (µg/m3) is depicted for both 2019 and 2020 (**Figures 4** and 5) [29].

The UV Aerosol Index concentration measured as an index value in (−2.0 to 0.5) is given in the below chart diagram for 2018–2022. Between March and April 2020, it's visible in the graph with the plummeting of UV Aerosol Index concentration values near the duration of 100 days and rising again with the lifting of lockdown that occurred phase-wise (**Figure 6**).

The CO concentration measured in (mol/m2) is given in the chart diagram below for 2018–2022. Between March and April 2020, it's visible in the graph with the plummeting of CO concentration values near the duration of 100 days and rising again with the lifting of lockdown that occurred phase-wise. The TROPOMI Explorer App provides a lucid visualization of different trace gas combinations; in this case, it's for CO concentration which is depicted as a 9-day mean vertically integrated column (**Figure 7**).

The HCHO concentration measured in (µmol/m^2) is given in the below chart diagram for 2018–2022. Between March and April 2020, it's visible in the graph with

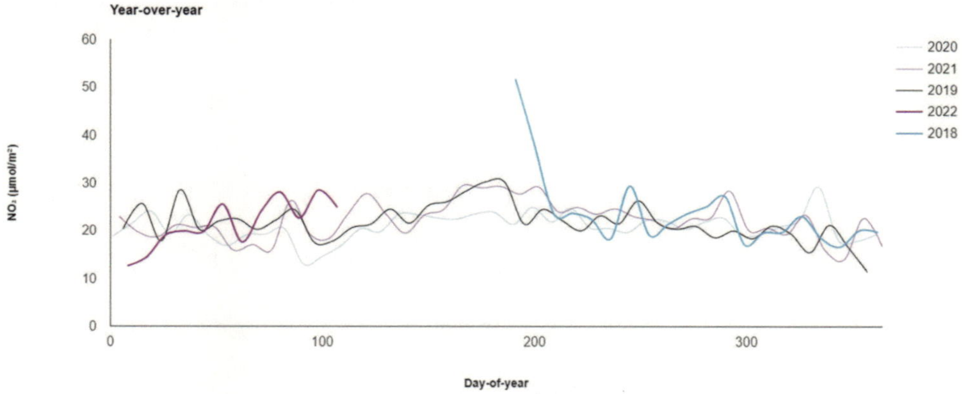

Figure 5.
The NO_2 concentration measured in (µmol/m^2) is given in the above chart diagram for 2018–2022. Between march and April 2020, it's visible in the graph with the plummeting of NO_2 concentration values near the duration of 100 days and rising again with the lifting of lockdown that occurred phase-wise.

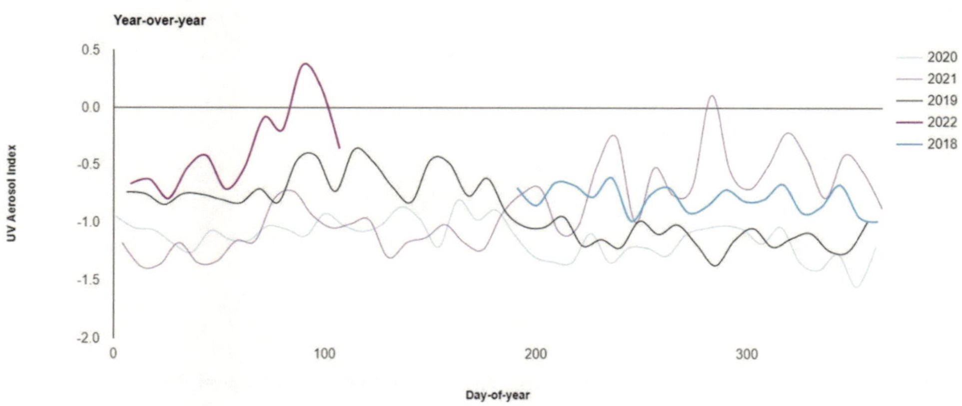

Figure 6.
The measured UV aerosol index concentration is given in the above chart diagram for 2018–2022.

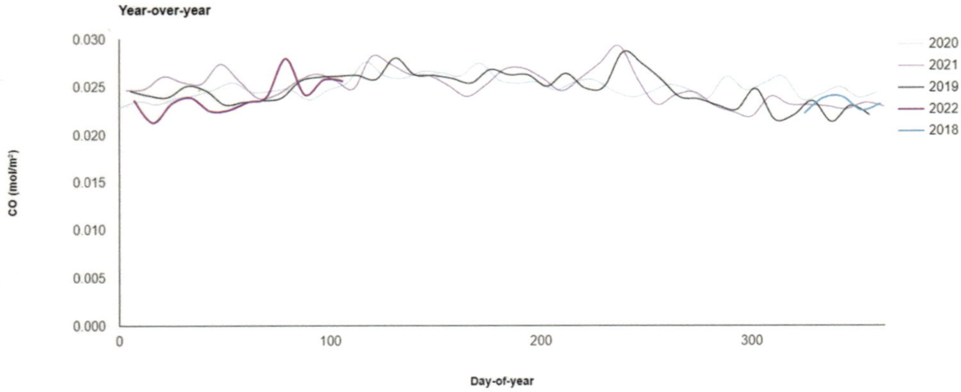

Figure 7.
The CO concentration measured in the above chart diagram for 2018–2022.

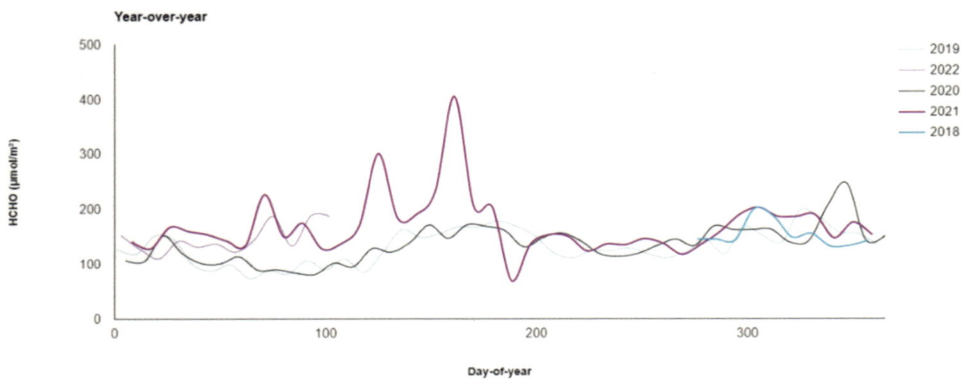

Figure 8.
The HCHO concentration measured in the above chart diagram for 2018–2022.

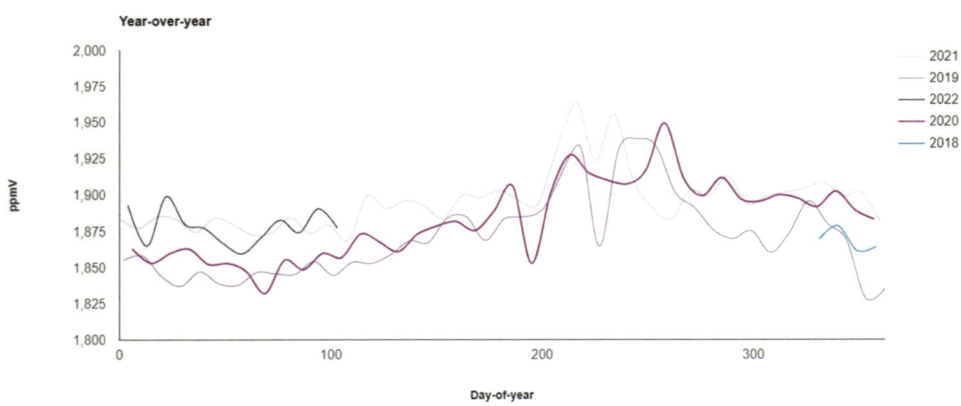

Figure 9.
The CH_4 concentration measured in (ppmV) is given in the above chart diagram for 2018–2022.

the plummeting of HCHO concentration values near the duration of 100 days and rising again with the lifting of lockdown that occurred phase-wise (**Figure 8**).

The CH_4 concentration measured in (ppmV) is given in the below chart diagram for 2018–2022. Between March and April 2020, it's visible in the graph with the

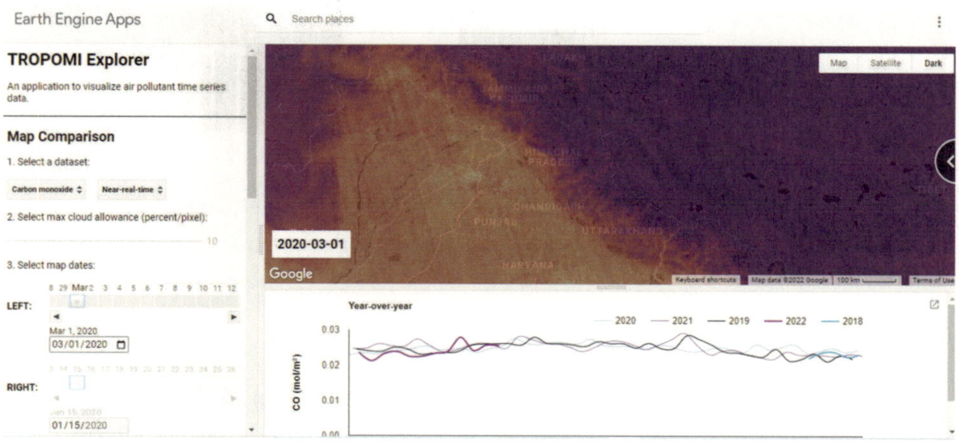

Figure 10.
The figure depicts the TROPOMI explorer (an application to visualize air pollutant time series data).

plummeting of CH_4 concentration values near the duration of 100 days and rising again with the lifting of lockdown that occurred phase-wise (**Figures 9** and **10**).

5. Conclusion

This paper discusses the results and discussions we showcase as a part of our study. In this paper, we have analyzed the impact of COVID-19 lockdown measures on air quality around the region of pan India, explicitly focusing on the state of Himachal Pradesh. These were based on the understanding and the observations of several trace gases from the Sentinel-5P. Sentinel-5P provides daily, global observations of multiple trace gases. The measured vertical column amounts are driven by emissions, including atmospheric, chemical source, and destination processes with multi-dimensional changes. We analyzed the time series of trace gas measurements from pan India to the regional level (state of Himachal Pradesh) and made a calculative comparison. We looked into the regional impacts of COVID-19 lockdown measures on the ambient air quality and anthropogenic emissions.

Furthermore, for the first time, we used a combination of five trace gases observed by Sentinel-5P, specifically NO_2, SO_2, CO, HCHO, and CH_4, to assess the impact of COVID-19-related lockdown measures on the trace gas concentrations levels. TROPOMI data have been used to analyze the implications of COVID-19 lockdown measures on the ambient air quality and air pollution levels. These studies have been based on NO_2 observations more than the other trace gases. We contemplate that the combined use of available trace gases from TROPOMI and the high spatial resolution of the sensor platform has massive potential for a significantly improved sector-specific analysis of the impact of the COVID-19 lockdown measures than previously encountered.

Author contributions

AG conceptualized, initiated, and managed this manuscript with contributions from CP. AG carried out a formal analysis. AG provided data curation and software support for Sentinel-5P data products with work on Google Earth Engine (GEE). AG prepared, edited, and co-managed the manuscript with minor changes done by CP.

Conflict of interest

The authors who contributed to this research work declare no conflict of interest.

Data availability

The Operational versions of all Copernicus Sentinel 5-P Data TROPOMI data are freely available from the European Union/ESA/Copernicus Sentinel-5P Pre-Operations Data Hub (https://s5phub.copernicus.eu; S5P Pre-Ops Data Hub, 2021) and these are also available on Google Earth Engine platform (https://developers.google.com/earth-engine/datasets/catalog/sentinel-5p). The administrative areas can be downloaded on (https://www.diva-gis.org/gdata).

Author details

Abhinav Galodha[1*], Chander Prakash[2] and Devansh Raniwala[3]

1 School of Interdisciplinary Research (SIRe), Indian Institute of Technology, Delhi (IIT Delhi), India

2 Department of Civil Engineering, NIT Hamirpur, Himachal Pradesh, India

3 The Neerja Modi School, Jaipur, Rajasthan, India

*Address all correspondence to: srz208737@sire.iitd.ac.in

References

[1] Aljazeera: Coronavirus in India: What We know About World's most Extensive Lockdown. 2020a. pp. 912-13 https://www.aljazeera.com/news/2020/05/india-coronavirus-crisis-200519120521747.html. [Accessed: June 17, 2020]

[2] Afshari R. Indoor air quality and severity of COVID-19: Where communicable and non-communicable preventive measures meet. Asia Pacific Journal of Medical Toxicology. 2020;**9**(1):1-2

[3] Dattakiran J. Impact of Lockdown on India's Electricity Sector, Energy A. 2020 http://www.energy-a.eu/impact-of-ongoing lockdown-on-Indias-electricity-sector-an-overview/. [Accessed: June 9, 2020]

[4] De Smedt I, Bai J, de Leeuw G, Theys N, Van Roozendael M, Sogacheva L, et al. Variations and photochemical transformations of atmospheric constituents in North China. Atmospheric Environment. 2018;**189**:213-226

[5] BBC: India Extends Coronavirus Lockdown by Two Weeks. 2020a https://www.bbc.com/news/world-asia-india-52698828. [Accessed: June 17, 2020]

[6] Fioletov V, McLinden CA, Griffin D, Theys N, Loyola DG, Hedelt P, et al. Anthropogenic and volcanic point source SO 2 emissions derived from TROPOMI on board Sentinel-5 precursor: First results. Atmospheric Chemistry and Physics. 2020;**20**(9):5591-5607

[7] Granier C, Darras S, van der Gon HD, Jana D, Elguindi N, Bo G, et al. The Copernicus Atmosphere Monitoring Service global and regional emissions (April 2019 version). [Research Report] Copernicus Atmosphere Monitoring Service; 2019

[8] Ialongo I, Virta H, Eskes H, Hovila J, Douros J. Comparison of TROPOMI/Sentinel-5 precursor NO 2 observations with ground-based measurements in Helsinki. Atmospheric Measurement Techniques. 2020;**13**(1):205-218

[9] Jain S, Sharma T. Social and travel lockdown impact considering coronavirus disease (COVID-19) on air quality in megacities of India: Present benefits, future challenges, and ways forward. Aerosol and Air Quality Research. 2020;**20**:1222-1236. DOI: 10.4209/aaqr.2020.04.0171

[10] Prabhjote G. The most congested cities in India lie vacant amid the nationwide lockdown. Business Insider India, https://www.businessinsider.in/india/news/most-congested-cities-in-india-low-lie-vacant-midst-thenationwide-lockdown/articleshow/75243376.cms, last access: 30 March 2021. 2020

[11] Kharol SK, Fioletov V, McLinden CA, Shephard MW, Sioris CE, Li C, et al. Ceramic industry at Morbi as a significant source of SO2 emissions in India. Atmospheric Environment. 2019;**1189**:223. DOI: 10.1016/j.atmosenv.2019.117243

[12] Lerot C, Theys N, Volkamer R, Müller JF, Zarzana KJ, Kille N, et al. Wildfires enhance global nitrous acid emissions and levels of regional oxidants. Nature Geoscience. 2020;**13**(10):681-686

[13] Lokhandwala S, Gautam P. The indirect impact of COVID-19 on the environment: A brief study in the Indian

context. Environmental Research. 2020;**188**:109807

[14] Mahato S, Pal S, Ghosh KG. Effect of lockdown amid COVID-19 pandemic on air quality of the megacity Delhi, India. Science in the Total Environment. 2020;**730**:139086. DOI: 10.1016/j.scitotenv.2020.139086

[15] Van Geffen JH, Lorente A, Boersma KF, Eskes HJ, Veefkind JP, De Zeeuw MB, et al. Quantification of nitrogen oxide emissions from the build-up of pollution over Paris with TROPOMI. Scientific Reports. 2019;**9**(1):1-10

[16] Theys K, Hufsky F, Lamkiewicz K, Almeida A, Aouacheria A, Arighi C, et al. Computational strategies to combat COVID-19: Practical tools to accelerate SARS-CoV-2 and coronavirus research. Briefings in Bioinformatics. 2021;**22**(2):642-663

[17] Landgraf J, Fu D, Bowman KW, Worden HM, Natraj V, Worden JR, et al. High-resolution tropospheric carbon monoxide profiles were retrieved from CrIS and TROPOMI. Atmospheric Measurement Techniques. 2016;**9**(6):2567-2579

[18] Sun K, Zhu L, Cady-Pereira K, Chan Miller C, Chance K, Clarisse L, et al. A physics-based approach to oversample multi-satellite, multispecies observations to a standard grid. Atmospheric Measurement Techniques. 2018;**11**(12):6679-6701

[19] Kumari P, Toshniwal D. Impact of lockdown measures during COVID-19 on air quality–a case study of India. International Journal of Environmental Health Research. 2020;**32**(3):503-510. DOI: 10.1080/09603123.2020.1778646

[20] Levelt PF, Stein Zweers DC, Aben I, Bauwens M, Borsdorff T, De Smedt I, et al. Air quality impacts of COVID-19 lockdown measures were detected from space using high spatial resolution observations of multiple trace gases from sentinel-5P/TROPOMI. Atmospheric Chemistry and Physics Discussions, 1-53. 2021;**10**(7):356-368. DOI: 10.5194/acp-2021-534 in review, 2021

[21] Li C, McLinden C, Fioletov V, Krotkov N, Carn S, Joiner J, et al. India is overtaking China as the World's largest emitter of anthropogenic sulfur dioxide. Scientific Reports. 2017;**7**:14304. DOI: 10.1038/s41598-017-14639-8

[22] Li X, Jónsson S, Cao Y. Interseismic deformation from Sentinel-1 burst-overlap interferometry: Application to the southern Dead Sea fault. Geophysical Research Letters. 2021;**48**(16):e2021GL093481

[23] Ludewig A, Kleipool Q, Bartstra R, Landzaat R, Leloux J, Loots E, et al. In-flight calibration results of the TROPOMI payload on board the Sentinel-5 precursor satellite. Atmospheric Measurement Techniques. 2020;**13**(7):3561-3580

[24] Singh KD, Goel V, Kumar H, Gettleman J. India, Day 1: World's largest coronavirus lockdown begins. The New York Times, https://www.nytimes.com/2020/03/25/world/asia/india-lockdown-coronavirus.html, last access: 30 March 2021. 2020;**9**(1):312-321

[25] Verhoelst T, Compernolle S, Pinardi G, Lambert JC, Eskes HJ, Eichmann KU, et al. Ground-based validation of the Copernicus sentinel-5p TROPOMI NO_2 measurements with the NDACC ZSL-DOAS, MAX-DOAS, and Pandonia global networks. Atmospheric Measurement Techniques. 2021;**14**(1):481-510

[26] Sharma M, Jain S, Lamba BY. Epigrammatic study on the effect of

lockdown amid Covid-19 pandemic on air quality of most polluted cities of Rajasthan (India). Air Quality, Atmosphere & Health. 2020;**13**(10):197-1165

[27] Veefkind JP, Aben I, McMullan K, Förster H, De Vries J, Otter G, et al. TROPOMI on the ESA Sentinel-5 precursor: A GMES mission for global observations of the atmospheric composition for climate, air quality, and ozone layer applications. Remote Sensing of Environment. 2012;**120**:70-83

[28] Schneising O, Buchwitz M, Reuter M, Bovensmann H, Burrows JP, Borsdorff T, et al. A scientific algorithm to simultaneously retrieve carbon monoxide and methane from TROPOMI onboard Sentinel-5 precursor. Atmospheric Measurement Techniques. 2019;**12**(12):6771-6802

[29] Qiu Z, Ali MA, Nichol JE, Bilal M, Tiwari P, Habtemicheal BA, et al. Spatiotemporal investigations of multi-sensor air pollution data over Bangladesh during COVID-19 lockdown. Remote Sensing. 2021;**13**(5):877

Section 7

LULC Dynamic Modelling

Chapter 9

CA-Markov Approach in Dynamic Modelling of LULCC Using ESA CCI Products over Zambia

Charles Bwalya Chisanga, Chizumba C. Shepande and Edson Nkonde

Abstract

The Markov, Cell Atom and CA-Markov modules in TerrSet v19.0 have been applied to predict LULC maps for 2030 over Zambia. The European Space Agency Climate Change Initiative (ESA CCI) classified LULC maps for 2000, 2010 and 2020 were used in this study. The ESA-CCI LULC maps were reclassified using QGIS 3.20 into 10 classes. The 2000 and 2010 LULC maps were used to predict the 2020 LULC maps. The Kappa statistics between the 2020 reference and predicted LULC maps was kappa (0.9918). The probability and transition matrix between the 2010 and 2020 LULC maps were used as inputs into the CA-Markov module to generate the 2030 LULC map. The LULCC from 2020-2030 shows an expansion and contraction of different classes. However, Built-up (42.38% [481.82 km^2]) constitutes major changes among the LULC classes. However, Cropland, Dense forest, Grassland, Wetland and Bare land will reduce by 376.00, 1087.65, 70.60, 26.67 and 0.36 km^2, respectively. Other LULC changes from 2020-2030 are in seasonally flooded grassland (94.66 km^2), Sparse forest (497.05 km^2), Shrub land (410.11 km^2) and Water body (77.63 km^2). The prediction of future LULC from historical LULC using CA-Markov model plays a significant role in policy making and land use planning.

Keywords: CA-Markov, cellular automata, ESA CCI, LULC, Markov

1. Introduction

Land use/land cover (LULC) has been defined as the physical composition and characteristics of land elements on the earth's surface [1]. The changes in LULC are caused by both natural and anthropogenic factors such as deforestation, and the intensification of agriculture [1–3]. Leta et al. [2] noted that the changes in LULC involve multifaceted processes that are dynamic, non-linear human-nature interactions that cause substantial land surface changes. The world-wide changes in LULC trajectory in the recent past have been characterised by gains and losses in agriculture and forests, respectively [2–4]. Researchers like Leta et al. [2], Pérez-Vega et al. [4], and Kolb et al. [5] argue that LULC changes are associated with major changes in forest land to agriculture, built-up and deforestation.

The dynamics of LULC change may be monitored using remotely sensed data (satellite imagery). The satellite imagery data is provided at different spatial, temporal and spectral resolutions and this is used to detect changes on the earth's surface [6]. The prediction of LULC change is vital in strategic developmental plans and in the management of LULC [7, 8]. There are myriads of Land Use Land Cover Change (LUCC) models that have been developed and these includes; Markov model, Cellular Automata (CA) models, statistical models, GeoMod, evolutionary models, CLUE-S model, hybrid models and multi-agent models [9–13]. Markov models can quantitatively predict the dynamic changes in landscape patterns [13]. Unfortunately, these models cannot resolve the spatial patterns of landscape change [10, 14]. Conversely, the CA models can predict the spatial distribution of landscape patterns but fail to predict temporal changes [13]. Because of the above reasons, researchers use a combination of CA and Markov to model LULCC dynamically [10, 13]. The most convenient tool used in simulating the spatio-temporal LULCC and processes in the landscape is the CA-Markov model [11, 15, 16]. Researchers have used the CA-Markov model to model dynamically the spatio-temporal LULCC and predict future scenarios [15, 17–26].

Using a Markov process, the future state of a system can be simulated using the immediately preceding state [8]. The Markov model describes LULC change from one time period to another and this is the basis used to project future changes [8, 27]. The Markov outputs are the transition probability, area matrices and probability images of LULC change from the initial time to time two which displays the nature of change as the basis for projecting a future time period. The LULCC has been analysed in Europe, USA, South America, Asia and Africa [2, 28]. The expansion of agricultural fields in Africa has been influenced by population growth [2]. Agriculture has been recognised as the main driver of land use change (LUC). The dynamics of urban growth are linked to factors associated with demography especially in developing countries.

The future LULC condition can be determined using modelling such as CA-Markov models. The CA-Markov is a combination of cellular automata and Markov. It is a LULC prediction procedure that adds an element of spatial contiguity and the likely spatial distribution of transitions to Markov change analysis [10]. The CA-Markov models can be used to model and monitor LULC change at spatial and temporal scale [8]. It has been used by [27, 29–31] to successfully model future LULCC.

Researchers involved in LULCC agree that the intellectual foundation of validation for land use change models is insufficient [2]. The literature reviewed suggest insufficient research on rigorous validation of LUCC models has been conducted [32, 33]. The objectives of this study were to investigate and analyse the spatio-temporal ESA CCI LULCC for 2000, 2010 and 2020 and to predict LULCC for 2030. The study attempts to analyse the spatio-temporal LULC change based on 2000, 2010, 2020 and 2030 using Zambia as a case study. The study also simulated the spatial variation in LULC change in 2000, 2010 and 2020 and used CA-Markov model to predict 2030 LULCC.

2. Materials and methods

2.1 Study area

The study covers the whole country, Zambia is used in this as shown in **Figure 1**. It is located in southern Africa and its spatial extent is between −8.27° to −18.075°

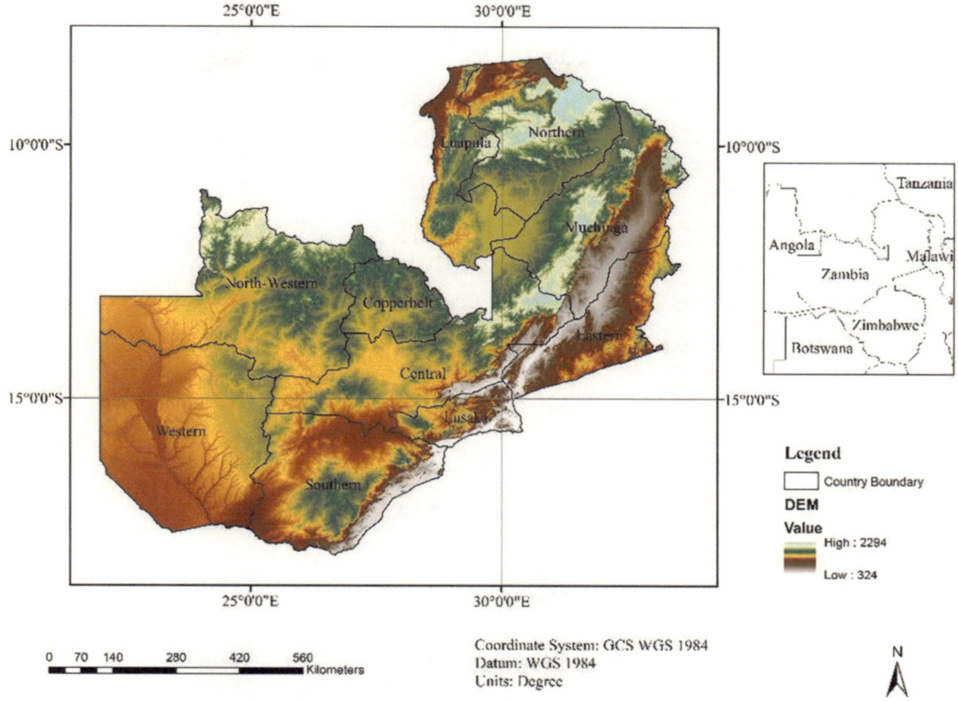

Figure 1.
Location of Zambia in southern Africa.

latitude South and 22.00° to 33.7° longitude East. Zambia is divided into three Agro-ecological Regions (AERI, AERII and AERIII) and ten provinces as shown in the figure below.

2.2 Sources of classified satellite imagery data

The global land cover maps utilised in this study were the European Space Agency Climate Change Initiative (ESA CCI) datasets [34, 35]. The LULC maps for 2000, 2010 and 2020 were downloaded from the ESA web site (https://cds.climate.copernicus.eu/cdsapp#!/dataset/satellite-land-cover?tab=form). The LULC maps for Zambia were clipped using the shapefiles for the administrative boundary. The annual ESA CCI LULC maps v2.1.1 (2000, 2010) and v2.0.7 (2020) have better quality in the classification and representation of change compared to v1.6.1 epoch-based datasets. The spatial resolutions of the LULC maps are at 300 m. These maps are reliable and cover the whole globe [34, 35]. The individual pixel value matches the labels of the land cover classes as defined by the United Nations (UN) Land Cover Classification System LCCS [35]. The UN LCCS classifiers having 22 classes makes provisions for further conversion into the Plant Functional Types distribution required by the Earth System Models. The global land cover datasets are produced in Coordinate Reference System (CRS).

2.3 Reclassification of land use types

The ESA CCI LULC datasets have 22 classes [34, 35] which were reclassified into 10 classes (**Table 1**). The reference for integration into 10 classes is the Food and

	Class name	2000	2010	2020
			Area in km^2	
1	Cropland	45733.401	46380.686	46055.166
2	Seasonally, flooded grassland	16348.384	16284.212	16369.564
3	Sparse forest	66244.035	65002.903	65425.752
4	Dense forest	444321.515	450001.335	449153.425
5	Shrub land	129878.716	123798.324	123985.97
6	Grassland	8333.316	8286.197	8216.149
7	Wetlands	27777.511	28531.536	28540.857
8	Built-up	401.316	642.154	1102.362
9	Bare land	41.93	45.207	44.571
10	Waterbody	13947.966	14055.535	14134.276
	Total	**753028.09**	**753028.09**	**753028.09**

Table 1.
Classified LULC map of 2000, 2010 and 2020 in square kilometres.

Agriculture (FAO) Openforis EarthMap and Vito Landcover viewer tools located at http://www.openforis.org/newwebsite/tools/earth-map.html and https://lcviewer.vito.be. The ESA 22+ classes were reclassified into 10 classes using *r.reclass* module in QGIS software.

2.4 Land use and land cover change using CA-Markov model

Researchers have used CA-Markov to monitor LULCC and predictions [12, 15, 17, 19–22, 25, 26]. The CA-Markov model in TerrSet v19.0 has been adopted in this study to obtain reliable results for Zambia. The 2030 LULC map was predicted based on the state of 2020 LULC. The CA-Markov model in TerrSet v19.0 uses a Markov, CellAtom (Cellular Automata) and CA-Markov modules [10, 11]. The CA-Markov model was used to calculate the amounts of change that may occur to some selected locations in the future [10, 11]. It analyses two qualitative LULC maps from different dates and produces a transition matrix, a transition areas matrix, and a set of conditional probability images [10]. The CA-Markov model is a stochastic process model that describes the probability of change from one date to another [2]. The transition probability would be the probability that a land cover type (pixels) at the initial time (t0) changes to another land cover type at the second time (t1). The changes in LULC between 2000 and 2010 were used to develop a transition probability, area matrices and probability images.

The Markov modules were used to generate the suitability transition images and change matrix. The cellular automata transition rule was implemented in TerrSet v19.0 using a combination of FILTER and RECLASS modules [10]. The CellAtom module was used to generate transition suitability images. The FILTER and RECLASS modules were used to create filters and reclassifying the LULC maps before implementing the cellular automata transition rules [10]. On the other hand, the CA-Markov module output was the 2020 predicted LULC map. The VALIDATE module was used to produce the kappa statistics using the 2020 reference LULC map and the

Number	Values	Strength of agreement
1	<0	Poor
2	0.01–0.40	Slight
3	0.41–0.60	Moderate
4	0.61–0.80	Substantial
5	0.81–1.00	Almost Perfect

Table 2.
Comparison and level of kappa agreement values [2, 37].

2020 predicted LULC map. More details on how the process was implemented are provided by Eastman [11].

2.5 Validation of the simulated map

An important stage in the development of any predictive change model is validation [10]. Validation is a procedure used to assess the quality of the predicted LULC map against a reference map [2]. The 2020 predicted LULC map was validated using the 2020 reference LULC map. The VALIDATE module in TerrSet v19.0 [11] was used as it provides a comparative analysis based on the Kappa Index of Agreement (KIA). KIA is essentially a statement of proportional accuracy, adjusted for chance agreement [36]. The kappa for stratum level location (KlocationStrata) is a quantification of the spatial accuracy within pre-identified strata, and it indicates how well the grid cells are situated within the strata [11]. The blend of Kstandard, Kno, Klocation, and Klocation strata scores is considered for a comprehensive evaluation of the overall accuracy both in terms of location and quantity. The statistics; AgreementQuantity, AgreementChance, AgreementGridCell, DisagreementGridCell, and Disagreement-Quantity are used to ascertain the strength of the agreement. The Kstandard, Kno, Klocation, KlocationStrata, AgreeGridcell and AgreeQuantity were used in validating the accuracy of the prediction LULC map. The possible ranges of map comparison and level of kappa agreement values are shown in **Table 2**.

3. Results and discussion

3.1 Gains and losses in LULC

The gains and losses by category are shown in **Table 3**. All LULC classes experienced gains and losses except Bare land and Built-up. Built-up and Bare land categories had gains and losses from 2000 to 2010, 2010–2020 and 2000–2020, respectively. The Built-up class from 2000 to 2020 increased by 697.62 km^2. From 2000 to 2020, the net gains in cropland, dense forest, wetland and water body were relatively large. Sparse forest, shrub land and grassland LULC were converted to other LULC classes. The sparse forest, shrub land and grassland LULC classes experienced net losses to other classes.

The LULCC from 2000 to 2020 in km^2 is shown in **Table 4**. Cropland, dense forest, wetlands, built-up, bare land and water body had an overall increase of 1.40%, 1.26%, 2.64%, 37.5, 7.25% and 0.77% from 2000 to 2010, respectively. However, there

Class name	Actual		Actual		Actual		Predictions	
	2000–2010		2010–2020		2000–2020		2020–2030	
	Gains	losses	Gains	losses	Gains	losses	Gains	losses
Cropland	1001.85	351.11	409.61	733.82	1388.32	1061.81	424.08	798.78
Seasonally, flooded grassland	197.81	262.67	272.80	188.04	464.71	444.81	282.57	188.50
Sparse forest	1257.64	2481.92	1497.22	1076.09	2710.59	3514.15	1599.30	1104.65
Dense forest	9205.64	3492.13	2377.90	3218.78	11426.64	6554.02	2453.63	3533.13
Shrub land	1597.37	7723.25	1763.11	1578.30	3234.90	9175.97	2016.56	1609.81
Grassland	84.58	131.75	37.41	107.61	120.78	238.16	38.51	109.27
Wetlands	995.02	247.65	140.41	131.10	1117.37	360.69	143.91	170.44
Built-up	239.82	0.00	457.80	0.00	697.62	0.00	479.36	0.00
Bare land	7.83	4.61	5.62	6.26	10.32	10.32	5.62	5.99
Waterbody	210.98	103.00	141.61	63.48	329.83	143.72	144.46	67.44

Table 3.
Gains and losses by category.

was an overall reduction of 0.39%, 1.91%, 4.91% and 0.57% in seasonally flooded grassland, sparse forest, shrub land, and grassland from 2000 to 2010, respectively. During 2010 and 2020, cropland, dense forest, grassland and bare land experienced a reduction of 0.71%, 0.19%, 0.85% and 1.43%, respectively. There was an overall increase in seasonally flooded grassland, sparse forest, shrub land, wetland, built-up and water body as shown in **Table 4**. Leta et al. [2] observed that the expansion of agricultural land for both domestic and commercial production are drivers of LULC change. Furthermore, cropland is on an increase while shrub land, forest and grassland are decreasing in the world.

From 2000 to 2020, cropland, dense forest, wetlands, built-up, bare land and water body had an overall increase of 3.28%, 1.92%, 8.27%, 56.98, 28.03% and 2.01%,

Class name	2000–2010	% Δ	2010–2020	% Δ	2000–2020	% Δ
Cropland	647.29	1.40	−325.52	−0.71	1550.23	3.28
Seasonally, flooded grassland	−64.17	−0.39	85.35	0.52	−72.99	−0.45
Sparse forest	−1241.13	−1.91	422.85	0.65	−2104.61	−3.28
Dense forest	5679.82	1.26	−847.91	−0.19	8689.64	1.92
Shrub land	−6080.39	−4.91	187.65	0.15	−11318.34	−9.55
Grassland	−47.12	−0.57	−70.05	−0.85	−81.00	−0.98
Wetlands	754.03	2.64	9.32	0.03	2503.33	8.27
Built-up	240.84	37.50	460.21	41.75	531.60	56.98
Bare land	3.28	7.25	−0.64	−1.43	16.33	28.03
Waterbody	107.57	0.77	78.74	0.56	285.81	2.01

Table 4.
LULCC from 2000 to 2020 in km².

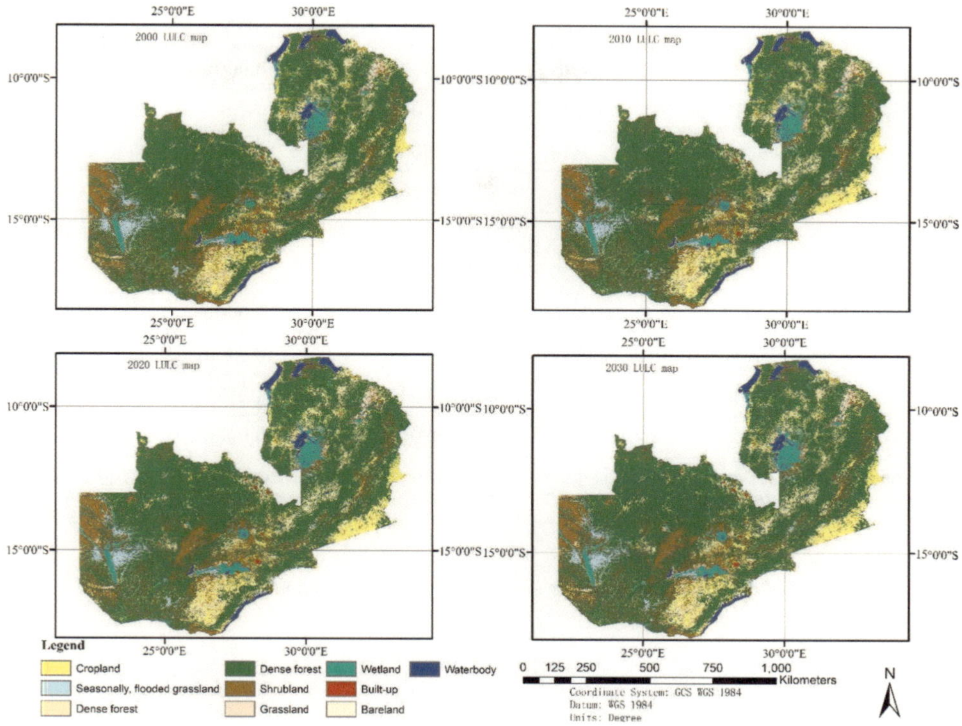

Figure 2.
Classified LULC for 2000, 2010, 2020 and 2030.

respectively. However, there was an overall reduction in seasonally flooded grassland, sparse forest, shrub land and grassland by 0.45%, 3.28%, 9.55% and 0.98%, respectively. The LULCC analysis of 2000 to 2010, 2010 to 2020 and 2000 to 2020, results indicated a positive increase in built-up and water body. Rapid population growth and migration of people from rural to urban areas have resulted in unprecedented changes in LULC. From 2000 to 2010, 2010–2020 and 2000–2020, built-up class had the largest increase at 37.50%, 41.75% and 56.98% followed by bare land at 7.25% (2000–2010) and 28.03% (2000–2020). Shrub land experienced a loss of 4.91% (2000–2010) and 9.55% (2000–2020) to other LULC classes. An increase in built-up and cropland LULC classes is anticipated to meet the demand of the population for residential and food production. This should be taken into consideration by policy makers and planners in future land use plans for sustainable management of natural resources such as soil fertility, water resources and forests (**Figure 2**).

3.2 Transition matrices for Markov 2020 and 2030 prediction

The transition matrix of Markov 2020 and 2030 predictions is shown in **Tables 5** and **6**. The CA-Markov model is simple to calibrate and can simulate LULCC dynamically with high efficiency [12]. Furthermore, it can simulate complex and multiple land cover patterns. The combination of Markov and Cellular Automata (CA-Markov) allows simulating the evolution of the geographical area represented by pixels. CA-Markov is a combined cellular automata/Markov change LULC prediction procedure that adds an element of spatial contiguity as well as knowledge of the likely spatial distribution of transitions to Markov change analysis [10]. Each pixel can take a value

	C 1	C 2	C 3	C 4	C 5	C 6	C 7	C 8	C 9	C 10
C 1	0.9824	0.0000	0.0000	0.0112	0.0060	0.0000	0.0000	0.0004	0.0000	0.0000
C 2	0.0000	0.9741	0.0000	0.0224	0.0034	0.0000	0.0000	0.0000	0.0000	0.0001
C 3	0.0000	0.0000	0.9527	0.0418	0.0051	0.0000	0.0000	0.0003	0.0000	0.0000
C 4	0.0026	0.0008	0.0036	0.9822	0.0056	0.0004	0.0046	0.0000	0.0000	0.0002
C 5	0.0043	0.0003	0.0050	0.0572	0.9311	0.0001	0.0001	0.0017	0.0000	0.0003
C 6	0.0017	0.0002	0.0002	0.0055	0.0113	0.9742	0.0000	0.0031	0.0000	0.0038
C 7	0.0000	0.0000	0.0000	0.0086	0.0010	0.0000	0.9811	0.0000	0.0000	0.0093
C 8	0.0011	0.0011	0.0011	0.0011	0.0011	0.0011	0.0011	0.9900	0.0011	0.0011
C 9	0.0095	0.0000	0.0024	0.0000	0.0000	0.0000	0.0000	0.1071	0.8810	0.0000
C 10	0.0001	0.0000	0.0002	0.0010	0.0018	0.0001	0.0141	0.0000	0.0000	0.9826

C 1 = Cropland; C 2 = Seasonally & flooded grassland; C 3 = Sparse forest; C 4 = Dense forest; C 5 = Shrub land; C 6 = Grassland; C 7 = Wetland; C 8 = Built-up; C 9 = Bare land; C 10 = Waterbody.

Table 5.
Transition matrix of Markov 2020 prediction based on LULC maps 2000 and 2010.

	C 1	C 2	C 3	C 4	C 5	C 6	C 7	C 8	C 9	C 10
C 1	0.9743	0.0000	0.0000	0.0137	0.0102	0.0000	0.0000	0.0017	0.0000	0.0000
C 2	0.0000	0.9786	0.0000	0.0190	0.0023	0.0000	0.0000	0.0001	0.0000	0.0000
C 3	0.0000	0.0000	0.9735	0.0218	0.0038	0.0000	0.0000	0.0009	0.0000	0.0000
C 4	0.0016	0.0012	0.0070	0.9829	0.0064	0.0001	0.0005	0.0001	0.0000	0.0001
C 5	0.0013	0.0006	0.0027	0.0127	0.9774	0.0002	0.0000	0.0047	0.0000	0.0005
C 6	0.0020	0.0001	0.0002	0.0018	0.0148	0.9770	0.0000	0.0039	0.0000	0.0003
C 7	0.0000	0.0000	0.0000	0.0040	0.0013	0.0000	0.9854	0.0000	0.0000	0.0092
C 8	0.0011	0.0011	0.0011	0.0011	0.0011	0.0011	0.0011	0.9900	0.0011	0.0011
C 9	0.0087	0.0000	0.0000	0.0022	0.0130	0.0000	0.0000	0.1238	0.8523	0.0000
C 10	0.0006	0.0000	0.0000	0.0004	0.0021	0.0004	0.0109	0.0000	0.0000	0.9855

C 1 = Cropland; C 2 = Seasonally & flooded grassland; C 3 = Sparse forest; C 4 = Dense forest; C 5 = Shrub land; C 6 = Grassland; C 7 = Wetland; C 8 = Built-up; C 9 = Bare land; C 10 = Waterbody.

Table 6.
Transition matrix of Markov 2030 prediction based on LULC maps 2010 and 2020.

from a finite set of states. All pixels are affected by a transition function that takes as arguments of the measured values and values of the neighbouring pixels as a function of time [8, 10, 38].

3.3 Validation of CA-Markov LULC prediction results

Table 7 shows the statistics of model validation. The Kappa Index of Agreement (KIA), Kno (kappa for no information), Klocation (kappa for location), Kstandard (kappa for standard), and KlocationStrata (kappa for stratum-level location) [39]

	Filter 3 x 3	Filter 5 x 5	Filter 7 x 7
Kstandard	0.9918	0.9918	0.9917
Kno	0.9939	0.9939	0.9938
Klocation	0.9932	0.9932	0.9932
KlocationStrata	0.9932	0.9932	0.9932
kappa	0.9918	0.9918	0.9917
AgreeGridcell	0.6707	0.6706	0.6706
AgreeStrata	0.000	0.000	0.0000
AgreeQuantity	0.2329	0.2329	0.2329
AgreeChance	0.0909	0.0909	0.0909
DisagreeQuantity	0.0010	0.0010	0.0010
DisagreeStrata	0.0000	0.0000	0.0000
DisagreeGridcell	0.0046	0.0046	0.0046

Table 7.
Kappa statistics for 2020 reference versus 2020 predicted LULC map.

shown in **Table 7** indicate the accuracy of the prediction. The Kstandard, Kno, Klocation, KlocationStrata, AgreeGridcell, AgreeStrata, AgreeQuantity, AgreeChange, DisagreeQuantity, DisagreeStrata and DisagreeGridcell using user-defined filters of 3x3, 5x5 and 7.7 shown in **Table 7** indicate the performance of the prediction. The Kno indicates the agreement between the 2020 reference and 2020 predicted LULC map. The Klocation was almost perfect and indicated the spatial accuracy in the overall LULC in each category between the predicted and reference map [40]. **Table 8** shows the comparison between the predicted and reference LULC map. The overall Kappa

	Class name	2020 reference	2020 predicted	Difference between reference and predicted	
			Filter5x5	Area km^2	%
1	Cropland	46055.166	47283.635	−1228.5	−0.16
2	Seasonally, flooded grassland	16369.564	16275.393	94.171	0.01
3	Sparse forest	65425.752	64139.428	1286.32	0.17
4	Dense forest	449153.425	453011.151	−3857.7	−0.51
5	Shrub land	123985.97	118560.381	5425.59	0.72
6	Grassland	8216.149	8252.317	−36.168	0.00
7	Wetlands	28540.857	30280.842	−1740	−0.23
8	Built-up	1102.362	932.914	169.448	0.02
9	Bare land	44.571	58.258	−13.687	0.00
10	Waterbody	14134.276	14233.771	−99.495	−0.01
	Total	**753028.092**	**753028.092**		

Table 8.
Comparison of 2020 reference and 2020 predicted LULC map.

between the 2020 reference and 2020 predicted map was 0.9918. The results indicate that reference and predicted LULC maps have many similarities. The difference between the maps is due to changes that will take place in the future LULC maps.

3.4 Predicted LULCC in 2030

Figure 2 shows the 2000, 2010, 2020 and 2030 LULCC maps over Zambia. The Built-up constitutes major changes among the LULC classes with an increasing trend at 42.38% (481.82 km^2) from 2020 to 2030 (**Table 8**, **Figure 3**). Researchers such as Jain et al. [41] have observed that built-up areas have increased while forests have decreased. The growth in built-up consequently causes a reduction in cropland or agricultural land. The 2020 to 2030 LULC prediction maps indicate that Cropland, Dense forest, Grassland, Wetland and Bare land will reduce by 0.82% (376.00 km^2), 0.24% (1087.65 km^2), 0.86% (70.60 km^2), 0.09% (26.67 km^2) and 0.80% (0.36 km^2), respectively. However, LULC classes of seasonally flooded grassland (0.58%, 94.66 km^2), Sparse forest (0.76%, 497.05 km^2), Shrub land (0.33%, 410.11 km^2), Built-up (42.38%, 481.82 km^2) and Water body (0.55%, 77.63 km^2) will increase from 2020 to 2030. Analysing historical LULCC from future LULCC using CA-Markov plays a significant role in forest management and land use planning [29, 31]. The CA-Markov model can be applied in modelling and quantifying transition rates and different states of diverse land uses [24]. The CA-Markov model can simulate larger areas efficiently [13] such has been undertaken in this study. However, CA-Markov is very sensitive in simulating smaller regions as noted by [13, 42].

The gains and losses by category indicate that Built-up will have gains of 479.36 km^2 between 2020 and 2030. The Cropland, Dense forest, Grassland, Wetland and Bare land LULC categories will experience greater losses compared to gains as shown in **Table 9**. On the other hand, seasonally flooded grassland, Sparse forest, Shrub land and Water body LULC categories will experience greater gains than losses. Therefore, LULC prediction is important as it contributes to national, and regional planning and management of natural resources [3].

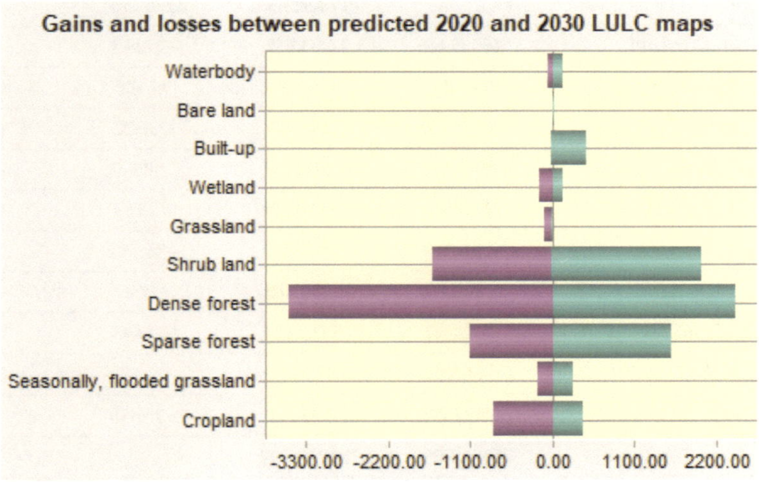

Figure 3.
Gains and losses by categories of LULC from 2020 to 2030 over Zambia.

	Actual			Predicted									
Class	**2000**	**2010**	**2020**	**2020**	**2030**	**2000–2010**	**% Δ**	**2010–2020**	**% Δ**	**2000–2020**	**% Δ**	**2020–2030**	**% Δ**
C 1	45733.40	46380.69	46055.17	46421.85	46045.85	647.28	1.40	−325.52	−0.71	321.76	0.70	−376.00	−0.82
C 2	16348.38	16284.21	16369.56	16280.12	16374.78	−64.17	−0.39	85.35	0.52	21.18	0.13	94.66	0.58
C 3	66244.04	65002.90	65425.75	65019.72	65516.76	−1241.13	−1.91	422.85	0.65	−818.28	−1.25	497.05	0.76
C 4	444321.51	450001.34	449153.42	450072.03	448984.38	5679.82	1.26	−847.91	−0.19	4831.91	1.08	−1087.65	−0.24
C 5	129878.72	123798.32	123985.97	123621.05	124031.17	−6080.39	−4.91	187.65	0.15	−5892.75	−4.75	410.11	0.33
C 6	8333.32	8286.20	8216.15	8285.54	8214.93	−47.12	−0.57	−70.05	−0.85	−117.17	−1.43	−70.60	−0.86
C 7	27777.51	28531.54	28540.86	28562.67	28536.00	754.03	2.64	9.32	0.03	763.35	2.67	−26.67	−0.09
C 8	401.32	642.15	1102.36	655.01	1136.83	240.84	37.50	460.21	41.75	701.05	63.59	481.82	42.38
C 9	41.93	45.21	44.57	45.30	44.94	3.28	7.25	−0.64	−1.43	2.64	5.93	−0.36	−0.80
C 10	13947.97	14055.54	14134.28	14064.80	14142.43	107.57	0.77	78.74	0.56	186.31	1.32	77.63	0.55
Total	**753028.09**	**753028.09**	**753028.09**	**753028.09**	**753028.09**								

C 1 = Cropland; C 2 = Seasonally & flooded grassland; C 3 = Sparse forest; C 4 = Dense forest; C 5 = Shrub land; C 6 = Grassland; C 7 = Wetland; C 8 = Built-up; C 9 = Bare land; C 10 = Waterbody.

Table 9.
Percent changes in LULC classes from 2000 to 2030.

4. Conclusions

The prediction of future LULC from historical LULC using CA-Markov model plays a significant role in policy making, land use planning and in natural resource management. There are major changes in built-up areas during the historical (2000 and 2020) and future (2030). An increase in built-up and cropland LULC classes is anticipated to meet the demand of the population for residential and food production. This should be taken into consideration by policy makers and planners in future land use plans for sustainable management of natural resources. This study has shown that the ESA CCI products can be used in LULC analysis and in predicting future LULCC.

Acknowledgements

The authors would like to thank ESA for the provision of the ESA CCI LULC maps and codes for converting NetCDF to Gtiff used in this study.

Author contributions

Downloading, preparation and analysis of the ESA CCI data by CBC. Methodology development by CBC. Model setup, simulation and validation by CBC. Drafting of the original manuscript by CBC. Abstract and conclusion by CCS. Reviews and editing by EN and CCS.

Contribution to the field statement

The information generated from change detection can be used in land use planning.

The European Space Agency Climate Change Initiative (ESA CCI) products can be used in LULCC analysis and in predicting future LULCC at 300 m spatial resolution.

The prediction of future LULC from historical LULC using CA-Markov model plays a significant role in policy making and land use planning.

Funding information

There was no funding for this research.

Competing interests

The authors declare no competing interests.

Data availability statement

The datasets analysed during the current study are available in the ESA CCI repository (https://cds.climate.copernicus.eu/cdsapp#!/dataset/satellite-land-cover?

tab=form). The reclassified datasets are available from the corresponding author on reasonable request.

Institutional Review Board Statement

Not applicable.

Informed Consent Statement

Not applicable.

Author details

Charles Bwalya Chisanga[1]*, Chizumba C. Shepande[2] and Edson Nkonde[3]

1 Department of Plant and Environmental Sciences, Copperbelt University, Kitwe, Zambia

2 University of Zambia, Lusaka, Zambia

3 Zambia Meteorological Department, Ministry of Transport and Communication, Lusaka, Zambia

*Address all correspondence to: cbchisanga@gmail.com

References

[1] Yulianto F, Prasasti I, Pasaribu JM, Fitriana HL, Zylshal Haryani NS, Sofan P. The dynamics of land use/land cover change modeling and their implication for the flood damage assessment in the Tondano watershed, North Sulawesi Indonesia, Modeling Earth Systems and Environment. 2006;**2**:20. DOI:10.1007/s40808-016-0100-3

[2] Leta MK, Demissie TA, Tränckner J. Modeling and Prediction of Land Use Land Cover Change Dynamics Based on Land Change Modeler (LCM) in Nashe Watershed, Upper Blue Nile Basin, Ethiopia. Sustainability. 2021;**13**:17. DOI: 10.3390/su13073740

[3] Saputra MH, Lee HS. Prediction of land use and land cover changes for north sumatra, indonesia, using an artificial-neural-network-based cellular automaton. Sustainability. 2019;**11**:1-16. DOI: 10.3390/su11113024

[4] Pérez-Vega A, Mas JF, Ligmann-Zielinska A. Comparing two approaches to land use/cover change modeling and their implications for the assessment of biodiversity loss in a deciduous tropical forest. Environmental Modeling Software. 2012;**29**:11-23

[5] Kolb M, Mas JF, Galicia L. Evaluating drivers of land-use change and transition potential models in a complex landscape in Southern Mexico. International Journal of Geographical Information Science. 2013;**27**:1804-1827

[6] Halmy MWA, Gessler PE, Hicke JA, Salem BB. Land use/land cover change detection and prediction in the north-western coastal desert of Egypt using Markov-CA. Applied Geography. 2015; **63**:101-112. DOI: 10.1016/j.apgeog.2015.06.015

[7] Ibrahim WYW, Ludin ANM. Spatiotemporal Land Use Change Analysis Using Open-source GIS and Web Based Application. International Journal of Built Environment Sustainability. 2014;**2**:10. DOI: 10.11113/ijbes.v2.n2.64

[8] Nguyen TTH, Ngo TTP. Land use/land cover change prediction in Dak Nong Province based on remote sensing and Markov Chain Model and Cellular Automata. Journal of Vietnamese Environment. 2018;**9**:132-140. DOI: 10.13141/jve.vol9.no3.pp132-140

[9] Dawn CP, Steven MM, Marco AJ, Matthew JH, Peter D. Multi-agent systems for the simulation of land-use and land-cover change: A review. Annals of the Association of American Geographers. 2016;**93**:314-337. DOI: 10.1111/1467-8306.9302004

[10] Eastman JR. IDRISI Taiga: Guide to GIS and Image Processing Volume-Manual version. 16.02. Worcester, MA, USA: Clark Labs, Clark University. 2009. p. 325

[11] Eastman JR. The IDRISI Applications Programming Interface for the TerrSet Geospatial Monitoring and Modeling System: A User's Guide. Version 4.0, Worcester, USA. 2019. p. 1–41

[12] Liping C, Yujun S, Saeed S. Monitoring and predicting land use and land cover changes using remote sensing and GIS techniques - A case study of a hilly area, Jiangle, China. PLoS One. 2018;**13**:1-23. DOI: 10.1371/journal.pone.0200493

[13] Yang C, Wu G, Chen J, Li Q, Ding K, Wang G, Zhang C. Simulating and forecasting spatio-temporal characteristic of land-use/ cover change

with numerical model and remote sensing: A case study in Fuxian Lake Basin China, European Journal of Remote Sensing. 2019;**52**:374–384. DOI: 10.1080/22797254.2019.1611387

[14] Balzter H, Braun PW, Kohler W. Cellular automata models for vegetation dynamics. Ecological Modelling. 1998; **107**:113-125. DOI: 10.1016/S0304-3800 (97)00202-0

[15] Hamad R, Balzter H, Kolo K. Predicting Land Use/Land Cover Changes Using a CA-Markov Model under Two Different Scenarios. Sustainability. 2018;**10**:1-26. DOI: 10.3390/su10103421

[16] Eastman RJ. TerrSet Geospatial Monitoring and Modeling System - Manual., Worcester, USA: Clark Labs, Clark University. 2016. p. 395

[17] Abdulrahman AI, Ameen SA. Predicting Land use and land cover spatiotemporal changes utilizing CA-Markov model in Duhok district between 1999 and 2033. Academic Journal of Nawroz University. 2020;**9**:18. DOI: 10.25007/ajnu.v9n4a892

[18] Adhikari S, Southworth J. Simulating Forest Cover Changes of Bannerghatta National Park Based on a CA-Markov Model: A Remote Sensing Approach. Remote Sensing. 2012;**4**:3215-3243. DOI: 10.3390/rs4103215

[19] Chaula JA. Ca-Markov Model for Simulating Land use Land Cover Dynamics in Rufiji Delta of Tanzania. International Journal of Innovative Science Research Technology. 2019;**4**: 802-814

[20] Hua AK. Application of CA-Markov model and land use/land cover changes in Malacca river watershed Malaysia, Applied Ecology and Environmental Research. 2017;**15**:605–622. DOI: 10.15666/aeer/1504_605622

[21] Khawaldah HA, Farhan I, Alzboun NM. Simulation and prediction of land use and land cover change using GIS, remote sensing and CA-Markov model. Global Journal of Environmental Science Management. 2020;**6**:215-232. DOI: 10.22034/gjesm.2020.02.07

[22] Matlhodi B, Kenabatho PK, Parida BP, Maphanyane JG. Analysis of the future land use land cover changes in the gaborone dam catchment using ca-markov model: Implications on water resources. Remote Sensing. 2021;**13**:20. DOI: 10.3390/rs13132427

[23] Samat, N., 2009: Integrating GIS and CA-MARKOV model in evaluating urban spatial growth Malaysian. Journal of Environmental Management, 10, 83–100

[24] Sang L, Zhang C, Yang J, Zhu D, Yun W. Simulation of land use spatial pattern of towns and villages based on CA-Markov model. Mathematical and Computer Modelling. 2011;**54**:938-943. DOI: 10.1016/j.mcm.2010.11.019

[25] Sunday A, Abdulkadir A, Anene NC. Simulation and Prediction of Urbanization in Makurdi City Nigeria using CA-Markov Technique. FUTY Journal of the Environment. 2020;**14**: 12-21

[26] Tadese S, Soromessa T, Bekele T. Analysis of the Current and Future Prediction of Land Use/Land Cover Change UsingRemote Sensingand the CA-MarkovModel in Majang Forest Biosphere Reserves of Gambella. Southwestern Ethiopia Science World Journal. 2021;**2021**:18. DOI: 10.1155/ 2021/6685045

[27] Nguyen HTT, Pham TA, Doan MT, Tran PTX. Land use/land cover change

prediction using multi-temporal satellite imagery and multi-layer perceptron markov model. International Archives of Photogrammetry Remote Sensing and Spatial Information Sciences ISPRS Archives. 2020;**54**:99-105. DOI: 10.5194/isprs-archives-XLIV-3-W1-2020-99-2020

[28] Faichia C, Tong Z, Zhang J, Liu X, Kazuva E, Ullah K, et al. Using RS Data-Based CA–Markov Model for Dynamic Simulation of Historical and Future LUCC in Vientiane, Laos. Sustainability. 2020;**12**:1-20. DOI: 10.3390/su12208410

[29] Behera MD, Borate SN, Panda SN, Behera PR, Roy PS. Modelling and analyzing the watershed dynamics using Cellular Automata (CA)-Markov model - A geo-information based approach. Journal of Earth System Science. 2012; **121**:1011-1024. DOI: 10.1007/s12040-012-0207-5

[30] Chen C-F et al. Multi-Decadal Mangrove Forest Change Detection and Prediction in Honduras, Central America, with Landsat Imagery and a Markov Chain Model. Remote Sensing. 2013;**5**:6408-6426. DOI: 10.3390/rs5126408

[31] Yirsaw E, Wu W, Shi X, Temesgen H, Bekele B. Land Use/Land Cover change modeling and the prediction of subsequent changes in ecosystem service values in a coastal area of China, the Su-Xi-Chang region. Sustainability. 2017;**9**:1-17. DOI: 10.3390/su9071204

[32] Pontius RG, Schneider LC. Land-cover change model validation by an ROC method for the Ipswich watershed, Massachusetts, USA Agriculture Ecosystem Environment. 2001;**85**:239–248, DOI:10.1016/S0167-8809(01) 00187-6

[33] Pontius RG, Huffaker D, Denman K. Useful techniques of validation for spatially explicit land-change models. Ecological Modelling. 2004;**179**:445-461. DOI: 10.1016/j.ecolmodel.2004.05.010

[34] ESA. ESA Quick user guide of the Land Cover State products in GTiff and NetCDF formats. UCL-Geomatics, Belgium: ESA CCI partnership. 2015. p. 2

[35] European Space Agency. Land Cover CCI Product User Guide Version 2. Tech. Rep, UCL-Geomatics, Belgium: ESA CCI partnership. 2017. p. 105

[36] Pontius RG Jr. Quantification error versus location error in comparison of categorical maps. Photogramm. Eng. Remote Sensing. 2000;**66**:1011-1016

[37] Viera AJ, Garrett JM. Understanding interobserver agreement: The kappa statistic. Family Medicine. 2005;**37**:360-363

[38] Quintero GV, Moreno RS, García MP, Guerrero FV, Alvarez CP, Alvarez AP. Detection and projection of forest changes by using the markov chain model and cellular automata. Sustainability. 2016;**8**:1-13

[39] Mosammam HM, Nia JT, Khani H, Teymouri A, Kazemi M. Monitoring land use change and measuring urban srawl based on its spatial forms: The case of Qom city Egyptian Journal of Remote Sensing and Space Science. 2016;**20**:103-116. DOI: 10.1016/j.ejrs.2016.08.002

[40] Nadoushan MA, Soffianian A, Alebrahim A. Predicting urban expansion in arak metropolitan area using two land change models. World Applied Sciences Journal. 2012;**18**:1124-1132

[41] Jain S, Siddiqui A, Tiwari PS, Shashi M. Urban Growth Assessment using CA Markov Model: A case study of Dehradun

City. In: 9th International Geographic Union. Delhi, India. 2016. p. 9

[42] Berling S, Wu J. Modeling urban landscape dynamics: A case study in Phoenix USA. Urban Ecosystem. 2004;7: 215-240. DOI: 10.1023/B: UECO.0000044037.23965.45

Section 8

Landscape Vulnerability for Planning Hydroelectric Projects

Chapter 10

Strategically Planning of Hydroelectric Projects for Reduce the Physical Vulnerability of Landscape in Upper Sutlej Valley, Western Himalayas, India

Amit Kumar Jamwal and Vikram Sharma

Abstract

Hydropower's development in the Himalayas region is major concern because area is prone to the geo hazards. The high vulnerability of physical landscape pays attention on the planning of hydropower's projects. This qualitative empirical research in western Himalayas, present the vulnerability of region and impacts of hydroelectric projects on physical landscape. The IPCC frame work of vulnerability assessment was used to assess the vulnerability in upper Sutlej valley. The indicator based methodology and Geographic information System (GIS) & Remote Sensing (RS) applications were adopted to highlight the impacts and vulnerability. The strategic buffering equal distance analysis was done and this indicates the ignorance of hydropower planning processes. The GIS mapping indicated the excessive development of hydroelectric projects in single river valley and strategic planning emphasizes to follow aerial equidistance between two projects while introducing hydropower projects in the Sutlej valley or any other valleys of the Himalayas region. The suggested strategy shall control the physical, social and economic losses in study region. In addition, this strategy will work as guidelines to develop hydropower projects in other valleys of the Himalaya.

Keywords: Himalayas, hydropower's, hazards, vulnerability

1. Introduction

Hydropower project not leads to environmental air, soil, water pollution and believed as the source of green energy. But in recent time hydropower's development become the major concern in Himalayas region because of hazards vulnerability and its impacts [1] Upper Valley of Sutlej River is known for its major hydroelectric projects (25) of different phases and categories which included the 4 projects were commissioned, 4 projects were under construction, 9 were of under commissioned and 6 were of under investigation [2] (**Figure 1**). The upper valley of River Satluj was highly affected because of landslides incidences. These incidences were common in

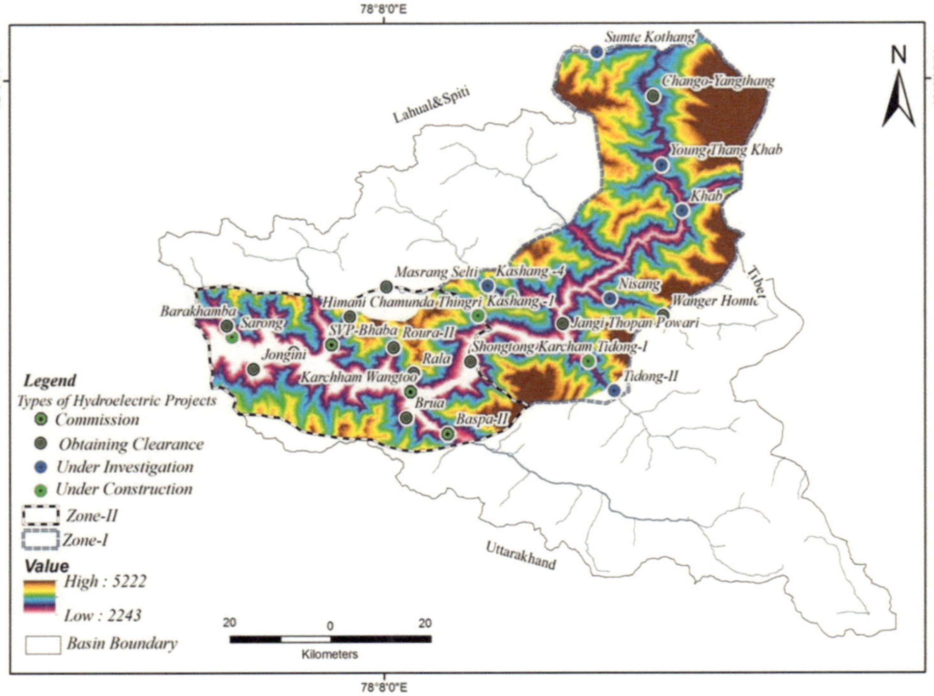

Figure 1.
Location of hydroelectric projects in Kinnaur district, upper Satluj Valley.

Himalaya's region but here frequency of these incidences were very high, It was evident from the previous study that maximum landslides incidences were occurred because of its fragile topography and haphazard development of hydroelectric projects [3]. The active landslides (Sudha Rang Dakhu, Shongtong, Urni, Tapari) were noticed in the region because of the slope failure incidences [4].

The social issues were related to the livelihood, properties, horticulture and agriculture losses were also observed in the study region [5]. The main objectives of the research were to observe the physical vulnerability and develop strategic planning of hydroelectric projects development. Vulnerability assessment is basically interaction of exposure, susceptibility and resilience of system. The system exposure indicates the threatening from the environmental and manmade hazards. The susceptibility of system raises its weakness through environment exposure. The resilience of the system is the strategic, technical part of planning where procedures are adopted and suggested to make resilient the system. The indicator ranking based methodology is one of important method to assess the vulnerability of selected region [6]. The cumulative environmental impact assessment methodology was adopted in river valley of western Himalayas region. This includes the study of two important river valley Yamuna, Alaknanda and Satluj river valley. It was evident from the cumulative impact assessments reports that all river valleys have same environment issues (Degradation of physical landscapes and loss of livelihood options) which were results of haphazard hydropower's development [7]. The identified problems related to the high number of hydropower's project were found in one river valley without following the any equidistance and buffering of hydro projects. Secondary research also revealed that if we follow the strategy of equidistance and buffers of hydroelectric projects then we can control the physical degradation of landscape [6].

2. Study area

The study region of the upper Satluj valley in Himachal Pradesh was extend from 31^0 30′ 12” N to 32^0 22′16”N and 77^0 40′16″E to79^0 13′16″E and also covered the area of 6450km^2. The study region has high altitude which extended from 1200 m to 6705 m. The valley area was divided into two broad categories on the basis of climatic condition. The upper part of the Satluj valley had the climate types of semi-arid to arid temperate. Lower part of valley has high rainfall (816 mm) and the maximum precipitations (1278 mm) were occurred as a snowfall in upper part of valley. The study region is known for its complexities such as high dissection, low green cover (8%), high degree of slope (80% slope area > 30), high incidences of flash floods and landslides. The temperature of the upper part of valley is very low throughout the year but in winter season (December to February) is 8 degree C and can dip as low as −7 degree C. The highest temperature recorded in tapari was 300c in month of June and average high temperature was 15-200c. March to august. The region is known for its dry fruits (pine nuts; Pinus gerardiana) and apples (Alphonso, Kinnauri apple) orchards based livelihood.

3. Material and methods

Strategically planning of hydroelectric projects for avoiding the physical vulnerability of landscape in Sutlej valley was done on the basis of sensitivity (impacts) and selective indicators. The indicators were selected on the basis of their degree of sensitivity. Slope, aspect, lithology, soil texture, slope profile, relative relief, landslide, flood, earthquakes, social aspect indicators were taken. The indicators were classified into its subclasses and then the adverse impacts were observe, perceived and analyzed on these subclasses. The studied were conducted with in the buffer of 10 km form the main river [8]. The impacts of anthropogenic activities were studied within the buffer of 10 km from the main river. All indicators were studied with in the buffer of 10 km form the main river. The vulnerability index was generated for study area. The vulnerability indexes were analyzed on the basis of impacts and vulnerability relationship. The negative (1) and positive (2) relationship were analyzed on the basis of vulnerability formula (**Figure 2**) [9].

$$\text{Xij (N)} = \frac{\text{Max (Xij)–Xij}}{\text{Max (Xij)} - \text{Min(Xij)}} \tag{1}$$

$$\text{Xij(p)} = \frac{\text{Xij} - \text{Min(Xij)}}{\text{Max (Xij)} - \text{Min(Xij)}} \tag{2}$$

The geographic information system was used to analyze the sub classes. The GIS data based was generated from the Raster and vector data set. The qualitative and quantities values were normalized from 0 to 1. The software of ARC GIS, ERDAS were used to geo processed the raster and vector data set. Slope, slope aspect, slope profile, channel gradient, relative relief were analyzed on the basis of DEM (30 m). Soil texture, geological structure, Lithology map were prepared from the secondary map at the scale of 1:50,000. These parameters were also crossed verified in field with the help of GPS (Global Poisoning System).The different classes of land use land cover were studied on the basis of satellites images (ETM^+2019; 30 m). Earthquake

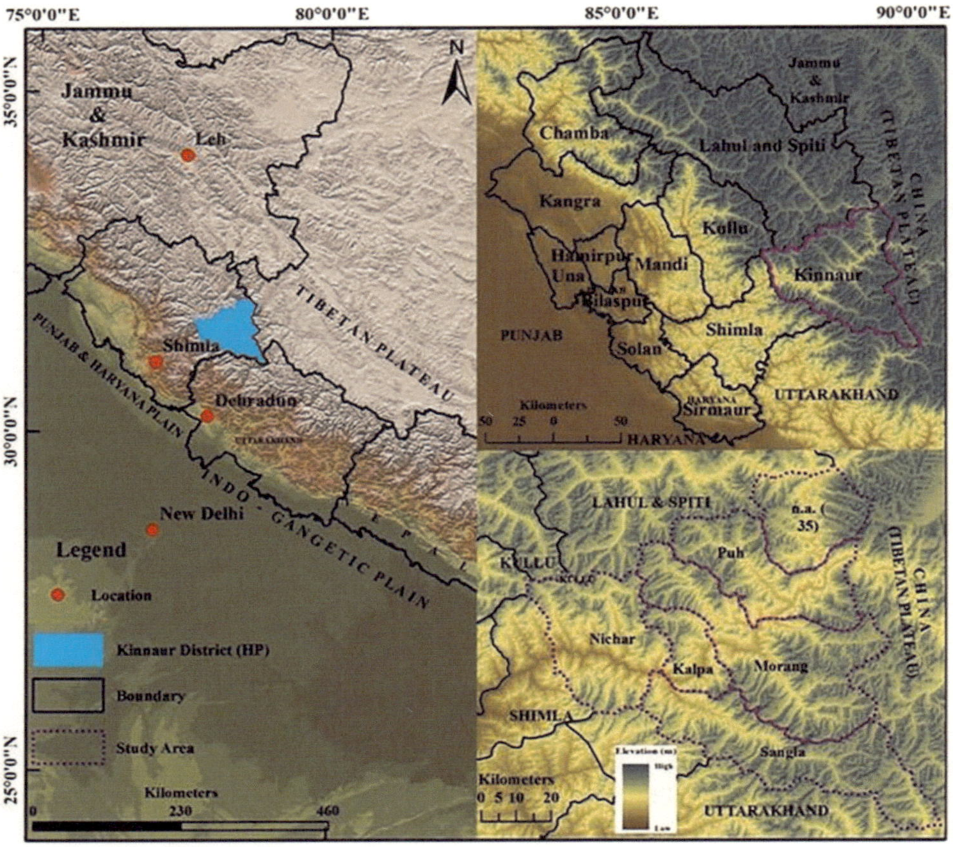

Figure 2.
Study area upper Satluj Valley district Kinnaur, Himachal Pradesh.

occurrence maps were prepared on the basis of secondary data (Earthquakes point data from 1901 to 2020) of incorporated research institution for seismology (IRIS) and SJVNL (2020).Rock fall, debris fall, debris slide, rock slide, rock creep and soil creep landslides were identified on the satellite images and ground observation were done with the help of GPS., the numbers of landslide incidences pixels were counted for different parameters such as slope, slope aspect, soil texture and land use to know the landslides status on subclasses of indicators. The comprehensive vulnerability index was used to analyze by using the IPCC, guidelines [10]. The flood incidences and their impact were analyzed through primary survey. The adaptability of the region was analyzed on the basis of adaptive indicators. The indictors were analyzed and quantified on the basis of people perceptions.

During the research period 2015–2020 the five consultative meetings, workshops were conducted in affected project area. The brainstorming, interactions, observations, perceptions were taken to analyzed the adaptive matrix of the region. During these meeting 67 respondents of specialized groups such as civil engineers, environmental engineers, lecturers, environmentalists research scholars and administrative were participated.

All possible layers of raster and vector were overplayed and sensitivity of the region was analyzed. The lithology map was extracted from the geological map of Himalayas at the scale of 1:100000 of geological survey of India (Government of India, 1989).

Soil texture was studied on the basis of field surveys and secondary data. The information was collected from the National Bureau of Soil Sciences (NBSS), and soil texture map was prepared at 1:50,000.

4. Result and discussion

$$\mathrm{VI} = \frac{\sum(\mathrm{Xij})\mathrm{p} + \sum(\mathrm{Xij})\mathrm{n}}{\mathrm{K}}$$

$$\text{Vulnerability in lower zone (VI)} = 7.5$$

4.1 Slope

The slope is one of the important indicators which indicated the sensitivity of physical loss. Higher slope degree results in rapid runoff and increased erosion rate (potential soil loss) with less ground water recharge potential [11]. The slope of the study area was classified into gentle, moderate, moderate steep, steep, very steep and vertical. The highest affected area was found under the steep, very steep and vertical. The vulnerability and physical landscape have positive relationship on the steep (0.51), very steep (0.51) and vertical slope (1). The negative relationship was found on the gentle (0) and moderate slope (0.34) which indicated the low sensitivity with low vulnerability (**Table 1**).

4.2 Slope aspect

The study region serve high portion of rocky land surface (98.7%) and sandy loam soil serve only 0.93%. The lowest portion was found under the sandy loam types of soil (0.2%). The southern aspect of slope had received high sensitivity (14.2) with positive vulnerability (1). The other aspect of the region such as east (0.75), South east (0.84), west (0.53) have high vulnerability. The northeast (0.31), east (0.75), south and west (0.53) have low vulnerability which evident the negative relationship. Pooh village, there were a large number of landslides occurred during summer-monsoon season due to snow melting. Soil creep, rock slide and rock fall were the common features [12].

4.3 Lithology

The rates of change of erosion, landslides are very much controlled by the lithology. The limestone topography is more vulnerable for construction with humid climatic condition. But it becomes more resist in the climate of arid to semi-arid [13]. The lithology consists the ortho quartzite, basic volcanic & limestone /dolomite was very sensitive (4.39) and highly affected which has high vulnerability (1) with positive vulnerability relationship (1). The high vulnerability was found under the topography namely (Pg3o) Boulder conglomerate sandstone, shale, clay (0.67), (Pt_{23}) Slate Phyllite, quartzite, gray shale (0.38), (Y) Granite & granitoid (0.91). The low vulnerability was recorded under the lithology types of (OC) Limestone, Siltstone (3.1) and (Pt3e) Greenish gray sandstone (1.79).

Indicators	Classification	Sensitivity (AV)	Relationship	Vulnerability (NV)	Rank relationship
1. Slope	Gentle (0°-5°)	0.24	(−)	0	$1^{(-)}$
	Moderate (5°-10°)	0.58	(−)	0.34	$2^{(-)}$
	Moderate Steep (10°-18°)	3.71	(+)	0.24	$4^{(+)}$
	Steep (18°-30°)	7.44	(+)	0.51	$3^{(+)}$
	Very steep (30°-45°)	7.53	(+)	0.51	$2^{(+)}$
	Vertical (45°-90°)	14.3	(+)	1	$1^{(+)}$
2. Slope aspect	Flat	0	(−)	0	$5^{(-)}$
	North	2.28	(−)	0.88	$3^{(-)}$
	North East	6.05	(−)	0.31	$4^{(-)}$
	East	10.9	(+)	0.75	$2^{(+)}$
	South East	12.1	(+)	0.84	$3^{(+)}$
	South	14.2	(+)	1	$1^{(+)}$
	South West	2.31	(−)	0.86	$2^{(-)}$
	West	7.95	(+)	0.53	$4^{(+)}$
	North West	0.69	(−)	0.95	$1^{(-)}$
3. Lithology	(Pt1) Regionally metamorphosed	5.61	(+)	1	2
	(Pt3e) Greenish gray sandstone	1.79	(−)	0.18	$2^{(-)}$
	(Y) Granite &granitoid	0.91	(+)	0	$4^{(+)}$
	(Pg3o) Boulder conglomerate sandstone, shale, clay	4.1	(+)	0.67	$2^{(+)}$
	(OC) Limestone, Siltstone, shale	3.1	(−)	0.53	$1^{(-)}$
	(Pt_{23}) Slate, Phyllite, quartzite, gray shale	2.7	(+)	0.38	$3^{(+)}$
	(Pt_2) Ortho quartzite, basic volcanic & limestone /dolomite	4.39	(+)	0.74	$1^{(+)}$
4. Soil texture	Coarse Texture	3.6	(−)	0.16	$2^{(-)}$
	Fine Texture	6.71	(+)	0.39	$2^{(+)}$
	Medium Texture	1.31	(−)	0	$1^{(-)}$
	Rocky / bad land	13.41	(+)	1	$1^{(+)}$
5. LULC	Settlements	0.21	(+)	0.0075	$7^{(+)}$
	Agricultural land	10.8	(+)	0.73	$4^{(+)}$
	Forest cover	10.2	(+)	0.69	$5^{(+)}$
	Barren / Wasteland	14.1	(+)	0.96	$2^{(+)}$
	Grass / grazing	14.6	(+)	1	$1^{(+)}$

Indicators	Classification	Sensitivity (AV)	Relationship	Vulnerability (NV)	Rank relationship
	Scrubland	1.46	(−)	0.093	$3^{(-)}$
	Water bodies	2.23	(−)	0.14	$6^{(-)}$
	Snow and glacier	0.1	(−)	0	$8^{(-)}$
6. Slope profile	Summital convexity	3.1	(+)	0.26	$4^{(+)}$
	Rectilinear section	2.5	(+)	0.42	$3^{(+)}$
	Free face	1.7	(−)	1	$2^{(-)}$
	Basal concavity	3.6	(+)	1	$1^{(+)}$
7. Relative relief	RR $_{4011}$	10.8	(+)	0.74	$2^{(+)}$
	RR $_{6014}$	3.2	(−)	0.24	$1^{(-)}$
8. River morphological	Channel Slope gradients	11.5	(+)	0	$2^{(+)}$
	Channel Slope gradients	15.7	(+)	1	$1^{(+)}$
9. Landslides	Large $_{(21)}$	21	(−)	1	2
	Landslides$_{(90)}$	90	(+)	1	1
10. Flood	1 Order streams	2	(−)	0	$4^{(-)}$
	2rd Order streams	8	(−)	0.5	$3^{(-)}$
	3rd Order streams	11	(+)	0.75	$2^{(+)}$
	4rd Order streams	14	(+)	1	$1^{(+)}$
11. Earthquakes	Faults	12.7	(+)	1	$1^{(+)}$
	MCT/Thrusts	11.4	(+)	8.3	$2^{(+)}$
	M ~ 6.1	6.4	(+)	0.58	$3^{(+)}$
	M ~ 5.5	4.2	(−)	0	$4^{(+)}$
12. Social issues	Livelihood losses	88	(+)	1	$1^{(-)}$
	Agriculture losses	71	(+)	0.6	$3^{(+)}$
	Horticulture losses	81	(+)	0.83	$2^{(+)}$
	Property losses	45	(+)	0	$4^{(+)}$

AV = actual Value NV = normalized value; MCT = main central Thrust; LULC = Land use land cover.

Table 1.
Vulnerability assessment index for the lower zone.

4.4 Soil texture

The soil texture of the region was classified in to coarse; fine, medium, rocky/bad land on the basis of secondary map and direct field observations. The soil of region was dry and consisting very fine particles and considered as the fine texture of soil. Such types of soil were found in the area like Tapri, Shongtong, Recongpeo, Powari, Kyari, Khwani, Pangi, Poh, Dakhu and Khab. The soil erosion is very common during the summer season (March to June). The rocky surface of the region was highly affected by the anthropogenic activities (blasting & tunneling) [14]. The Fine texture of the soil has high sensitivity (6.71) through the soil erosion and landslides. Fine texture (1.71) and Rocky/bad land (13.41) surface had high sensitivity and positive vulnerability score. However the medium texture (0.39) and coarse texture had less impact (3.6) with low vulnerability. Due to road and dam constructions, the rocky surface was highly influenced under the anthropogenic activities. Road cutting, tunneling and blasting were the major causes of the loss of physical landscape.

4.5 Land use and land cover (LULC)

Mass movement was largely influenced by land use and land cover change in the valley. More than 66.79 km^2areas was damaged due to landslides and construction activities, which include 10.8 km^2 agriculture, 10.2 km^2 forest, 14.1 km^2 wasteland, 14.6 km^2 grassland, 1.46 km^2 scrubland and 2.23 km^2 water bodies. The land degradation was high due to blasting, road widening and construction activities of hydropower. Settlements in an area of about 0.21 km^2 were damaged. The highest adversely affected area of 53.63 km^2 under LULC was affected in river valley. The highest adversely impacted area was occurred in case of wasteland (14.2 km^2). Snow covered area was found scarcely affected. The highest vulnerability was found under the grass/grazing land 14.6 (1), the waste land of valley was highly affected 2 (0.96) under the landslides and soil erosion. The scrubland had damaged area (1.46) 0.93 under the landslides and soil erosion but rarely affected with anthropogenic activities. Forest cover area of region had positive relationship with soil erosion and landslides [2]. The vulnerability score of the forest cover was 0.69 and ranked as 5. The agriculture area of the region was affected due to the landslides and soil erosion. The most of losses were found because of constructive activities. The water bodies (0.14), snow glacier areas (8) were also adversely affected.

4.6 Slope profile

Slope profile study was done on the basis of field survey. Free face segment and summital convexity of hill slopes were very common in the valleys of the Satluj valley, which cause high degradation of slopes and is also known as waxing slope. The Urni landslide slope was also affected due to the anthropogenic activities. Tanglling, Powari and Shongtong HEP area was also affected due to the construction activities.

The field surveys also made clear that most of the debris falls were found on this segment of slope. These types of slopes were found in the Satluj valley. A concave element of slope was very commonly found in lower portion of the Satluj valley. The free face, rectilinear section and summital convexity were affected mainly due to the construction activities. Summital convexity (3.1), Basal concavity (3.6) had high sensitivity which indicates the positive relation of impact and vulnerability. The free face (1.7) and rectilinear (2.5) were less affected by physical loss.

4.7 Relative relief (R_R)

GIS analysis indicated the high value of relative relief (R_R 5222 m). had the low vulnerability and indicates the negative value (0.24). Dissection index (DI = R_R/A_R) showed high value (0.64–0.93) indicating high vulnerability in terms of soil erosion, mass movement and flood in the Satluj valley [15]. The relative relief namely R_R 4011 m had high sensitivity which indicates the high vulnerability (7.4).

4.8 River morphological aspect

The river slope gradient was varies from 11.50 to 15.70 and the average channel slope gradient was 13.50. The high average slope gradient was one of the favorable conditions for the development of hydroelectric projects in this valley. The high slope gradient region had the highest number of allocated hydropower projects. High channel slope gradient is directly proposal to the high potential energy (Pe) and high kinetic energy (Ke). The high slope gradient region had the high sensitivity and vulnerability (1). The study region had the low drainage density (Dd = 0.32) and low texture of river. Higher the drainage density indicated the high surface runoff and high yield to soil erosion. But here low drainage texture and low value of bifurcation ratio (Rb 1.84) indicates that river channel had not strongly controlled by the parent rock. Because of that reason the region more vulnerable during the time of flash flood. The climate condition, vegetation covered, anthropogenic activities types factors had adverse impact on the river morphology. The valley was characteristic for a trellised drainage pattern. Here, stream pattern was turned towards the regional slope and geological structures like folds and faults.

4.9 Social issues

Livelihood, agriculture, horticulture and properties losses were major elements of social environment. The social issues were very sensitive in this region. The region has complex topography with adverse climatic condition. The agriculture and horticulture are main livelihood option in this region. The total geographic area was 624,000 hectare out of which 7.07 was cultivable land, 124,000 hectare was non agriculture and barren and uncultivable land was only 131.1200 hectare. The livelihood losses were commonly observed in the affected area under the constructed hydroelectric project [16]. The high sensitivity was found under the livelihood losses (1) and horticulture losses (0.83). The agriculture (0.6) and property losses (45) were ranked 3 and 4 respectively.

4.10 Landslide

The economic as well as loss of life due to landslides were considerably increased in the last century, and most of the landslides are due to global climate change, such as, El Niño and human activities [17]. Topographic variance (434–6448 m asml) influences precipitation pattern in the study region. Annual isohyets varied from 100 to 1400 mm. Rainfall decreased in this valley from the lesser to the Greater Himalaya. The lower part of valley of the Satluj valley falls under the sub-humid temperate to humid region. Here, rainfall pattern varied from 300 mm to 1400 mm. Total 111 landslides were identified out of which 21 were large (>0.198 km^2) and 90 were small (<0.198 km^2). The maximum numbers of landslides were noticed along the River Satluj and their validation was done through the GPS survey. Steep slopes, high relief, number of structural discontinuities and underlying geology were combined with anthropogenic activities which decrease the stability of slope [18]. Steep and vertical slopes have high degree of landslides. The highest affected area was found under the micro landslides, because these landslides were more in number. The constructed area of region was adversely affected under the anthropogenic activities which were ranked as 1. The area was affected under the large landslides ranked as 2 (**Table 1**).

4.11 Flood

Satluj valley is known for its divesting flood events the flood incidences were observed in 1993, 1995, 1997, 2000, 2005, 2007 and 2013. During the flash flood the water level was raise up to 15 $\sim$ 20 m from the normal water level and its water label go up to 10–12 time more than the normal water discharge. The main river of region consist the 4 type's stream; the first order (0) and second order stream (0.5) had the negative relation between flood and sensitivity. The **3 orders (0.75) a**nd 4 order streams (1) had the positive relationship of sensitivity [19]. It was made clear from the GIS study that affected area 139 km^2 due to flood was found and out of which 80 km^2 was found under high vulnerability.

4.12 Earthquake occurrences

The earthquakes occurrences were commonly observed in the study region. The 3.2 to 6.5 tremors of earthquakes were recorded in region. These earthquakes were sufficient to generate the landslides at micro and macro level. The study region have number of faults and MCT (main central thrust, separates the Greater Himalaya from the Lesser Himalaya) and faults were found in the lower zone where the Main Boundary Fault (MBF) passing through the region. However, sediment flux might be affected largely due to stronger earthquakes with greater magnitudes of MS > 7 [20]. Despite, seismically active mountain belts have shown link between earthquakes, landslides and fluvial sediment transport. The earthquakes sensitivity was analyzed on four deciding factors namely faults, MCT/Thrust, Magnitudes 6.1 and 5.5. The highest degraded areas were found near the tectonically sensitive area (faults & thrust). High vulnerability relationship were found under the all elements namely faults (1), MCT/Thrusts (8.3), 6.1(0.58) and 5.5 (0). The geological elements region was highly ranked under the vulnerability (1).

4.13 The peoples' perception survey

The peoples' perception survey was conducted in high altitude villages of valley like Apka (2454 m, Punag (1930 m) Tapri (2385 m) Agade (1599 m),Punasa (1791 m) and Surga (1591 m) (**Figure 3**). 67 households' survey samples were collected from the field. A structured questionnaire was designed and the direct survey was conducted by using the random sampling. The impacts of hydroelectric projects in the valley and the resultant socio-economic benefits of the communities were taken into account. About 46% of the total interviewees were perceived that the developmental footprints in terms of infrastructure, road, and colonies were visible in the valley. The local communities expressed their dissatisfaction in terms of their livelihood generation and inadequate compensation in lieu of their land loss. In the study area, 76% of the total interviewed individuals were perceived that agricultural loss, horticulture loss, drying up of springs and natural water sources, and impact of blasting on the individual houses were observed high in the valleys. People perceived that the earthquake incidences also increased in recent times and unscientific constructions, indiscriminate debris, dumping of muck along with the river banks result in many disorders, high adverse impacts in downslope regions. People were also believed that hydropower's project has bad impacts on their agriculture, horticulture, livelihood and social. People suggested that hydropower development should be controlled manner and should focus on tourism development as a sustainable livelihood option.

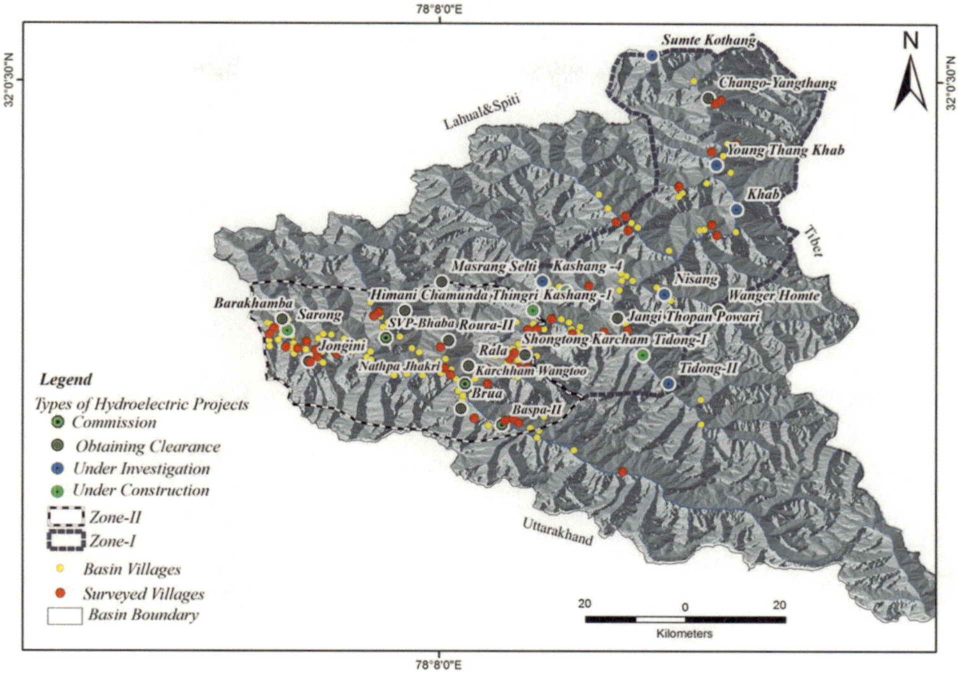

Figure 3.
Survey villages location in upper Satluj Valley, district Kinnaur.

Respondents also told that the tourism business also affected with hazards incidences if year is free from hazards incidences then tourist's number was high. But road were severely affected with the hazards like landslides, flood, anthropogenic activities (haphazard development of hydropower, blasting & tunneling).

4.14 Consultation meetings with the stakeholders

Regional Centre of Himachal Pradesh, Mohal-Kullu, Himachal Pradesh, India [3]. On the **Figure 4**.

The three strategic environmental assessment meetings were held in the deputy commissioner's office in Kinnaur, Rekong Peo on 26th Novem26th November 2014, 7th December 2015, and 23rd December 2016. The objective of these meetings is to interact with the local people and to know the environmental impacts on the social and physical environment. The project proponents, non-governmental organizations (NGOs), and local representatives were the main participants in these meetings. 83% of the respondents believed that their agricultural land was highly affected by the construction activities of hydropower. Local people admitted that the physical loss of landscape was high in the region. The local communities were against the large hydroelectric projects (>25 MW) construction in this valley.

The villagers of Sudharang Dakhu village explained that extensive tunneling and blasting were the major responsible causes for the continuous landslide and other physical loss. The people of the Satluj valley believed that the mitigation measure is very poor in and around hydroelectric projects. During the meeting, the participants replied that there is a need to recheck the inter distance of the hydroelectric project in the valley. The excessive development of hydroelectric projects in the signal valley

Figure 4.
A. Interaction with local people B. leakage of water from rock cracks C. degraded bank of Satluj River D. wide cross-section and gentle slope gradient of river channel E. landside of Sudharang Dakhu F. SEA meeting with government and local stakeholders G. slope cutting near Shongtong H. penstock of small hydroelectric project, near Tapri I. dry river bed of Satluj River.

should follow the inter distance. They suggested that there should be a fixed minimum radius between the two projects. 87.88% of participants suggested that there should be a fixed radius of 7 km between the two large projects, 87.88% suggested a 5 km radius between two small projects, and 81.82% and 3 km radius between two small projects. During these meetings and workshops, issues pertaining physical and social losses were discussed to a large extent. The stakeholders suggested that development of HEPs need to be controlled rationally so as to make them eco-friendly and sustainable (**Figure 5**).

The national park and wildlife sanctuary is an Eco-sensitive zone and within a radius of 10 km, all construction and anthropogenic activities are not permitted. This guideline was issued on 09.02.2011 regarding 'bird sanctuary' as a 'protected area' under Section 18 and 26 (A) of the Wildlife (Protection Act, 1972) [21–25]. Many wildlife sanctuaries and national parks are falls within the buffer of 10 km and this criterion was exploited by the project developer. Even all hydroelectric projects were not following any aerial distance criteria from one project to another (**Figures 6** and 7).

In this valley, there are different types of hydroelectric projects such as commissioned, under construction, and proposed projects. The 11 projects overlapped the area of other projects and not followed any inter-distance method. In the area of a high density of hydroelectric projects number area, the maximum physical losses and physical losses were recorded. The equidistance analysis of these hydroelectric projects was done and the geocentric mean was measured. The spatial analysis indicates that the maximum concentration of hydroelectric projects was found in the lower segments of the lower valley (**Figure 8**). It was evident from the field survey that the

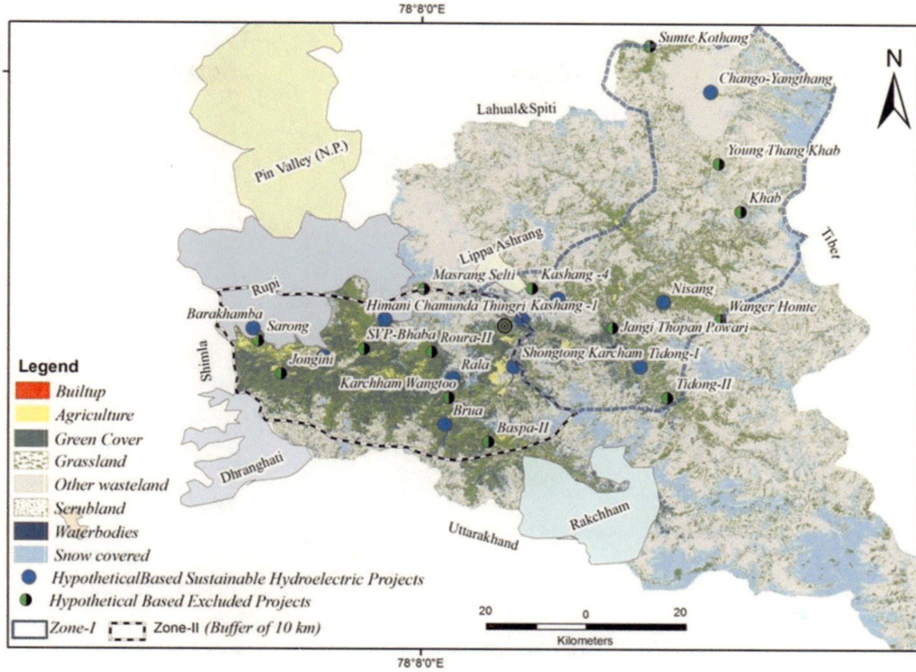

Figure 5.
Word life sanctuary within the buffer of 10 km.

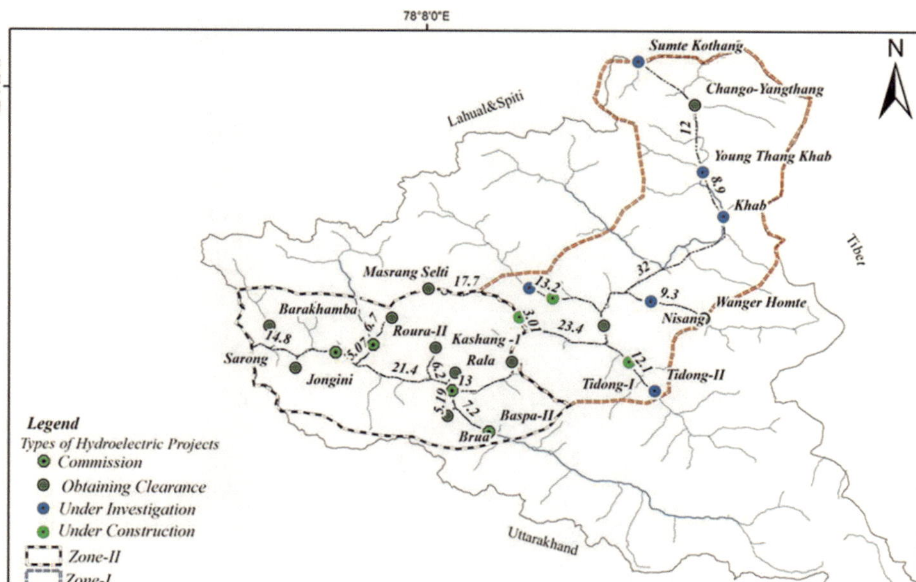

Figure 6.
Unequal distribution of hydroelectric project in Satluj Valley, district Kinnaur, Himachal Pradesh.

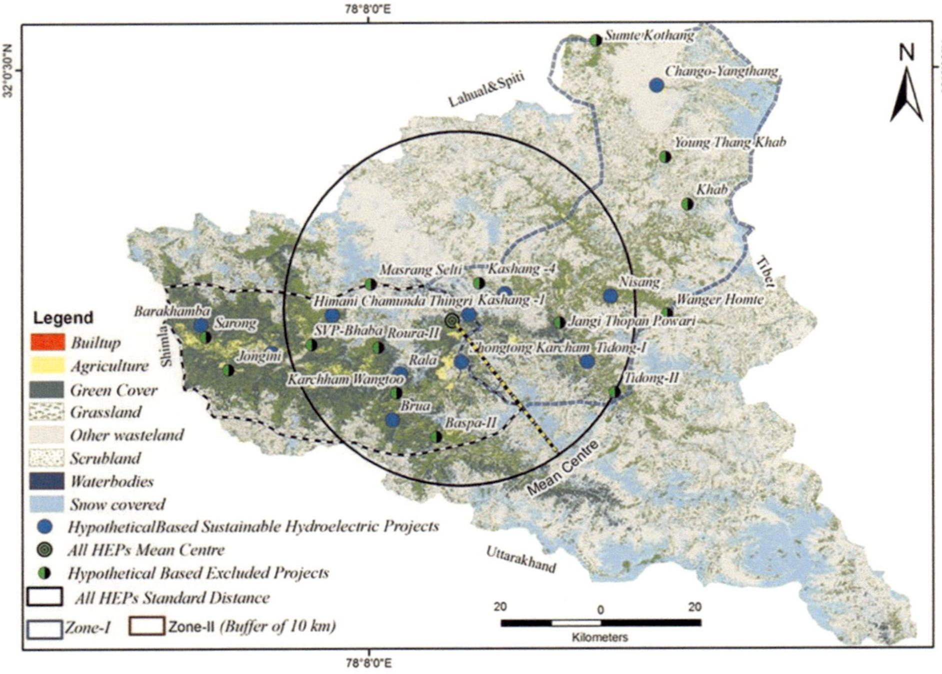

Figure 7.
The geocentric mean validate the high concentration of hydroelectric projects in particular area of valley.

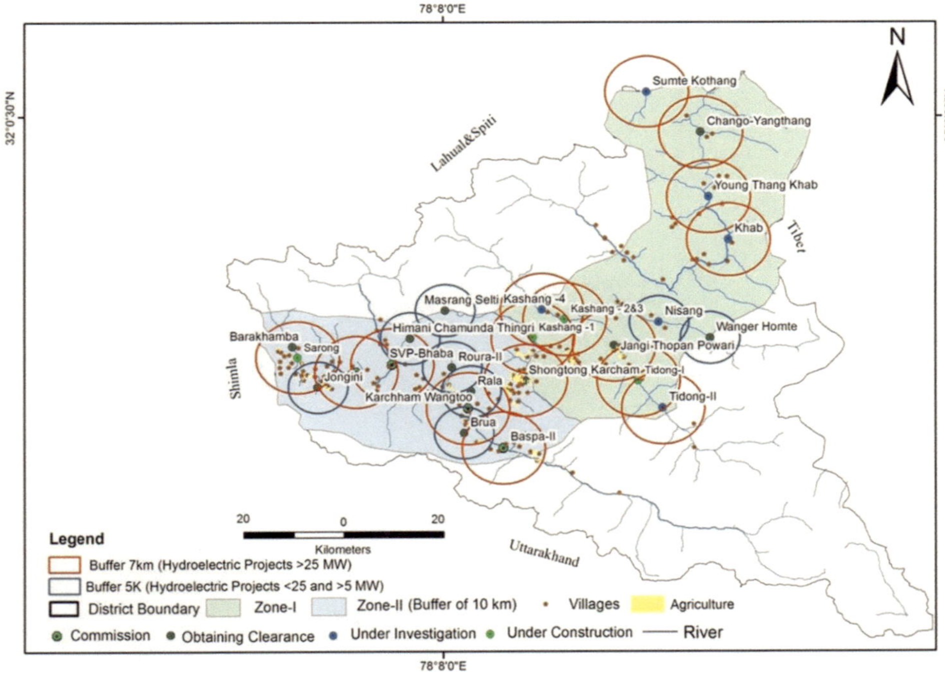

Figure 8.
Overlapping of hydroelectric projects.

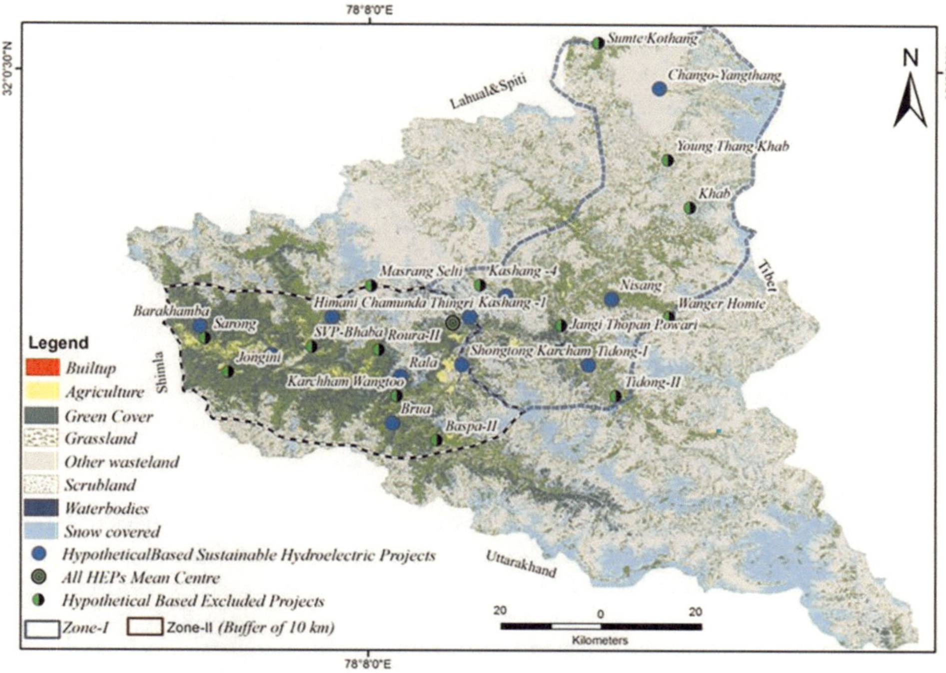

Figure 9.
Suggestive strategy to avoid the physical loss in study area.

maximum concentration of HEPs in a particular area was caused due to the physical degradation (**Figure 9**).

If we reduce the number of hydroelectric projects and develop the minimum equidistance criteria for the upcoming projects, the loss to physical landscape and resentment among the public regarding compensation and others would be minimum. The equidistance strategy can be implement in the River valley to avoid the physical loss the same model we can apply aerial inter distance between the projects where 7 km radius for large hydroelectric project 25 MW, 5 km for medium (5-25 MW) and 3 km should be suggested for small projects in mountain environment [3, 7]. Reservoirs / dams and powerhouse sites can be considered for a midpoint of equidistance buffer; however this can be depends on the nature of hydropower projects.

5. Conclusion

Strategic planning was done for avoiding the physical losses of valley. The physical losses were recorded in the valley. The indicators based analysis of the valley revealed that valley had the high sensitivity and vulnerability (7.5). The hydropower development was one of the important activities which responsible for the many environmental problems. It was evident from the GIS and remote sensing analysis that maximum losses were observed in the main river valley and near to the constructive activities. The anthropogenic activities were exorbitant the physical condition of the valley. The maximum numbers of landslides incidences were recorded near the constructive activities related to the hydro powers and roads. The incidences of slope failures, soil erosion, poor soil textures, low green cover (8%), high relative relief, floods, landslides, high anthropogenic activities indicates high vulnerability (7.5). The adverse impacts of hydropower projects were high as comparison to the positive impacts. The vulnerability of 753.41 km^2 area was identified under the physical loss which is very high in the valley and may exacerbate with the day-to-day increasing anthropogenic pressures. This was evident from the five interaction meeting that people of the study area were not satisfied with these developmental activities and they want control on the development activities. The brain storming meeting result revealed that there is need of equidistance between two hydroelectric projects. The agriculture losses, drying up of natural water resources, loss of horticulture land, muck deposition impact on river morphology, blasting impact on the people houses were noticed under the adverse impacts of hydroelectric projects.

The complex geophysical condition of the region is also responsible for the landslide incidences but the haphazardous developments of the hydroelectric projects aggravated the physical loss of the region. However, the maximum losses were found in and surrounding large hydroelectric projects and small hydroelectric projects region had the fewer incidences of physical degradation. To reduce the incidences of landslides there is a need to implement the fixed aerial equidistance from one project to the next project. However, there is a need to control hydropower development in the river valley in a sustainable way. If we succeed to implement the equidistance buffer between two projects then we succeed to control the environmental loss in and surrounding hydroelectric projects. This strategy would be beneficial in the mountain region to control the incidences of soil erosion, mass movement, and landslides.

Acknowledgements

The authors are heartily thankful to different stakeholders (local communities, project authorities and local government) for their constant support and cooperation during field study.

Author details

Amit Kumar Jamwal[1*] and Vikram Sharma[2]

1 Aryabhatta Geo-informatics and Space Application Centre (AGiSAC), Shimla, Himachal Pradesh, India

2 Department of Geography, Banaras Hindu University (BHU), Varanasi, Uttar Pradesh, India

*Address all correspondence to: amit.uprofft.feb2009@gmail.com

References

[1] Agarwal S, Kansal ML. Issues of hydropower development in Uttarakhand region of Indian Himalaya. Water and Energy International. 2017;**15**: 52-64

[2] Jamwal A, Kanwar N, Kuniyal JC. Use of geographic information system for the vulnerability assessment of landscape in upper Satluj valley of district Kinnaur, Himachal Pradesh, India. Geology Ecology and Landscapes. 2019;**4**(4):1-16. DOI: 10.1080/24749508.2019.160841

[3] Kuniyal JC, Lata R, Kumar A, Chand B, Kanwar N, Chaudhary S. Strategic environmental assessment of hydropower projects. Current Science. 2017;**113**(12):1239-2240

[4] Jamwal A, Sharma V. Vulnerability assessment of landslide with the help of geospatial approach in Western Himalayas, Upper Valley of river Sutlej, India, geospatial Technology for Environmental Hazards, advances in geographic information. Science. 2022; **18**:415-431. DOI: 10.1007/978-3-030-75197-5_18 4

[5] Kanwar N, Kuniyal JC, Kumar A, Nandi SK. Understanding climatic variability and Forest vulnerability due to hazards and anthropogenic activities: A study from the north-western Himalaya. India Journal of Himalayan Ecology Sustainable Development. 2017; **12**:45-56

[6] Kuniyal JC, Jamwal A, Kanwar N, Bhim C, Kumar K, Dhyani PP. Vulnerability assessment of the Satluj catchment for sustainable development of hydroelectric projects in the north-western Himalaya. Journal of Mountain Science. 2019;**16**(12):2714-2738. DOI: 10.1007/s11629-017-4653-z

[7] Kuniyal JC, Shashni S, Kumar A, Kanwar N, Chand B. Strategic environmental assessment. Current Science. 2015;**108**(4):480-481

[8] MoEF. EIA / EMP Report for Project / Activities Requiring Environmental Clearance under EIA Notification, 2006. India: Ministry of Environment, Forest and Climate Change; 2015 http://www.moef.gov.in/sites/default/files/final%20Booklet

[9] Birkmann J. Risk and vulnerability indicators at different scales, applicability, usefulness and policy implications. Environmental Hazards. 2011;**7891**:20-31

[10] Kumar M, Kalr N, Hukum S, Subrat S, Rawat PS, Singh RK, et al. Indicator-based vulnerability assessment of forest ecosystem in the Indian Western Himalayas: An analytical hierarchy process integrated approach. Ecological Indicators. 2021;**125**(107568): 1-15

[11] Griffiths DV, Fenton GA. Probabilistic slope stability analysis by finite elements. Journal of Geotechnical and Geo Environmental Engineering. 2004;**130**:507-518

[12] Marco C, Adriano R, Monica B. The slope aspect: A predisposing factor for land sliding? C. R. Geoscience. 2014;**30**:1-12. DOI: 10.1016/j.crte.2013.11.002

[13] Omar H, Bujang KH, Zenoddin B, Youssef M. Relationship between lithology factor and landslide occurrence based on information value (IV) and frequency ratio (FR) approaches, case study in north of Iran. Electronic Journal of Geotechnical Engineering. 2012;**17**: 79-89

[14] Sharma S, Kuniyal JC, Sharma JC. Assessment of man-made and natural hazards in the surroundings of hydropower projects under construction in the Beas Valley of north-western Himalaya. Journal of Mountain Science. 2007;**4**(3):221-236. DOI: 10.1007/s11629-007-0221-2

[15] Singh S. Geomorphology. 4th ed. Allahabad: Kalyan Publication; 2004a. pp. 381-382

[16] Lata R, Herojeet R, Dolma K. Environmental and social impact assessment: A study of hydroelectric power projects in Satluj Valley in district Kinnaur, Himachal Pradesh, India. International Journal of Earth Science and Engineering. 2017;**10**(02): 270-280. DOI: 10.21276/ijese.2017.10.0219

[17] Lee ML, Ng KY, Huang YF, Li WC. Rainfall-induced landslides in hula Kelang area, Malaysia. Natural Hazards. 2014;**70**:353-375

[18] Sharma S, Kuniyal JC. Hydropower development and policies in India: A case of Himachal Pradesh in the north-western Himalaya. India Energy Sources, Economic Planning and Policy. 2016; **11**(4):377-384. DOI: 10.1080/15567249.2011.633593

[19] Nity T, Parhi PK, Lohani AK, S. K. Chandniha analysis of precipitation variability over Satluj Valley, Himachal Pradesh, India: 1901–2013. Journal of Water and Climate Change. 2021;**12**(1): 127-135

[20] Hovius N, Meunier P, Lin CW. Prolonged seismically induced erosion and the mass balance of a large earthquake. Earth and Planetary Science Letters. 2011;**304**:347-355. DOI: 10.1016/j.epsl.2011.02.005

[21] NGT. Order on Okhla Bird Sanctuary. 2014. https://credai.org/assets/upload/judgements/resources/ngt–order-on-okhla-bird-sanctuary–3-4-2014.pdf [Accessed on 05 January 2022]

[22] Allen SK, Linsbauer A, Randhawa SS, Huggel C, Rana P, Kumari A. Glacial lake outburst flood risk in Himachal Pradesh, India: An integrative and anticipatory approach considering current and future threats. Natural Hazards. 2016;**14**:1-137. DOI: 10.1007/s11069-016-2511-x

[23] IPCC. Managing the risks of extreme events and disasters to advance climate change adaptation. In: Field CB et al., editors. A Special Report of Working Groups I and II of the Intergovernmental Panel on Climate Change. New York: Cambridge University Press; 2012. p. 582

[24] IPCC. Climate change 2014: Impacts, adaptation, and vulnerability. Part a: Global and sectorial aspects. In: Field CB et al., editors. Contribution of Working Group II to the Fifth Assessment Report of the Intergovernmental Panel on Climate Change. Cambridge, New York: Cambridge University Press; 2014. p. 1132

[25] Singh S. Geomorphology. 5th ed. Allahabad: Kalyan Publication; 2004b. pp. 267-296

Section 9

Identification of Groundwater Potential Zones

Chapter 11

Groundwater Potential Zone Identification Using GIS: Mettur, Salem District, Tamil Nadu

C. Prakasam and R. Saravanan

Abstract

Salem region is one of the drought inclined areas of Tamil Nadu. Being a rural zone, the area is finding a mitigation measure to keep the agribusiness on run. As a mitigation measure mapping the potential groundwater zones in the examination area will be a great alternate wellspring of water for the rural reason. The objective is to delineate the potential locations of groundwater zones using GIS software for the Metturtown panchayat, Salem district. The study area is inclined to drought. The different parameters that control the groundwater fluctuations, for example, geomorphology, contour, topography, LULC, soil; rainfall is broken down together as thematic maps. The slope map will be set up from DEM. These maps have been overlaid and the spatial analysis method, weighted overlay analysis using the GIS has been carried out. The ranking/weightage will be specified for every distinct bound of each thematic map. The weightages/rankings were relegated according to each thematic layer influence with respect to the Soil, LULC, and density of drainage, rainfall, and slope. The resultant maps display the groundwater potential zones ranging from very good to very poor zones.

Keywords: GIS, Salem, groundwater potential zones, drought

1. Introduction

The Salem locale of Tamil Nadu is an agro-economic area is pronounced as a drought inclined zone among 20 other regions in Tamil Nadu. Despite what might be expected these locales additionally had seemingly contradictory weather conditions: Tamil Nadu and Karnataka got ordinary rainfall, about 46%, recorded insufficient rainfall. Each of the three states additionally experienced far-reaching surges in August 2018 six locales in Tamil Nadu. Other than having wellsprings of water, the locale is inclined to dry season henceforth the mitigation measure ought to be in a way that utilizes these wellsprings of water in the examination area. On that note finding the zones where there is a potential wellspring of groundwater will be an imperative mitigation measure for drought zone or to a greater degree an adapt up strategy. The objective of this research work is to identify the groundwater potential location using the GIS approach in the project area.

Finding the groundwater zones in the field will be a dreary process; GIS applications make it simple in delineating the zones remotely and additionally give spatial and transient information about the groundwater zones in the locale. Magesh et al. [1] utilized the GIS strategies to delineate the groundwater potential location in Theni. The MIF procedures helped in delineating the potential zones in terms of scales from (very poor to very good). Waikar et al. [2] stated that the multi-criteria analysis using the GIS applications for hydro-geomorphological mapping and resource evaluation for water resource management is an effective analysis. The study was carried out to evaluate the geomorphic features and drainage patterns in the Parbhani District recommend the GIS application has appropriate methods for groundwater potential zone. Mwega et al. [3] stated Groundwater is a naturally available resource that sustains basic needs, farming, and industrial purposes. The generated groundwater potential zone is arranged into scales from high to low. Ramu et al. [4]. Some of the features that control groundwater zone mapping are soil, drainage density, LULC, topography, rainfall, geomorphology, incline, and contour. Rajendran [5]. Drought mitigation measures ought to be propelled on a war footing and in a sustainable way. Groundwater recharging ought to get top need and any dry season mitigation plans should incorporate this key part. Owolabi et al. [6]. Validation of the potential location of groundwater has been carried out using the borehole yield for the Buffalo catchment, South Africa. The study area is prone to water supply deficiency; hence the mapping technique came in handy. Doke et al. [7]. GIS-based methods result with precise accuracy and computation time is less than the traditional field methods. Jhariya et al. [8]. Low slope pays the way for low runoff and high infiltration. It also results in good groundwater recharge. Melese et al. [9]. Drainage density is inversely proportional to the groundwater probability. Saranya et al. [10]. The groundwater mapping help in future planning of the artificial recharge zones in the Kanchipuram district. The resultant shows that moderate potential is available for the study area.

The present study incorporates the GIS applications in delineating the groundwater potential zone in the Mettur panchayat town panchayat, Salem district.

2. Study area

For the study purpose, Mettur town (11°45′ N and 77°45′ E) of Salem district has been chosen. It has a spread area of 410 sq. km at an elevation of 292 m (958 feet), situated at 52 km north-west of Salem. The Mettur town has a population of 52,813. Mettur is known for its huge dam built-in 1934, which is still one of the best dams in the country. It also attracts tourists from all over India. The dam opens in June every year for irrigation in the Kaveri Delta. The River Cauvery divides the town into two parts as it traverses in between the Mettur (**Figure 1**).

Among 16 rainfall stations, the most extreme of 1073.32 mm records at Tholuthur (edges of the basin) and 1016.31 mm records at Mettur station inside the basin. The demand for water has increased in the study area. Paddy, Sugarcane, Maize, Turmeric, and Cotton are the main agricultural crops. It has 27 income towns and 14 panchayats. The soil type is red in situ soil is commonly occurring type. The woodland soil in timberland areas, dark cotton soil occurs in the north-eastern part and sandy-loamy blended soil in south-western area. The aquifer system ranges between 2.5 and 28 m with yields ranging from 10 to 840 lpm. The bore wells recorded transmissivity esteems ranging from 1 to 250 m^2/day.

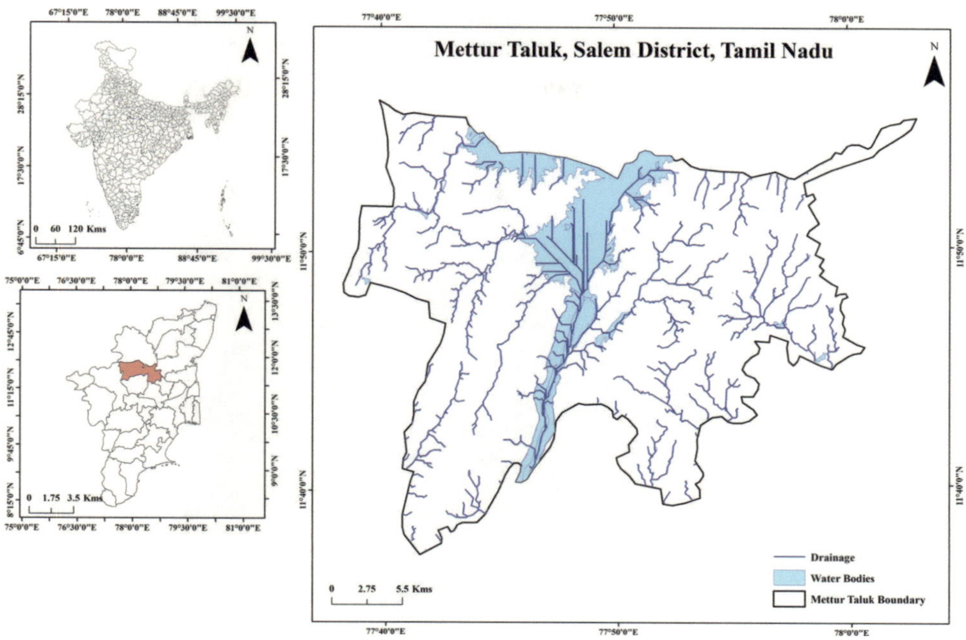

Figure 1.
Index map of the study area.

3. Methods and materials

There are six imperative indicators to be specific, (i) geology, (ii) slope, (iii) geomorphology, (iv) land use/cover, (v) drainage and (vi) lineament for groundwater prospects. These parameters are most commonly used parameters in identifying the potential location. These parameters have a direct relation to the groundwater level. Preparation of maps for these themes (with the exception of slope) in view of picture attributes, for example, tone, surface, shape, shading and association are standardized. Slope is gotten from ASTER DEM 30 m resolution. Thematic maps of the examination area were readied. To get a wholistic perspective of the previously mentioned indicators, overlay analysis is required. Task of rank to an individual class depended on the influence of these themes as announced in literature. Rank and weight-based thematic layers have been unified through GIS in order to find out the resultant groundwater potential zones. Overlay analysis was done from the determined multi-thematic layers in a GIS domain.

3.1 Slope

Higher level of slope leads to fast runoff. The slope map was developed from the ASTERDEM and classified into five categories (**Figure 2**). Slope values ranged from 0 to 90 degrees, as the study area contained hilly terrain. Gentle slopes (0–18) were spread over the area in a maximum range. Slopes greater than 54.1 degrees to 90 degrees represent the hilly slope in the region. 36% of the study area is having a hilly slope of 54.1–90 degrees and 64% of the study area is having a gentle slope. The degree of slope is inversely proportional to the groundwater potential. The weightage was according from 1 to 5 based on the degree of slope in the region.

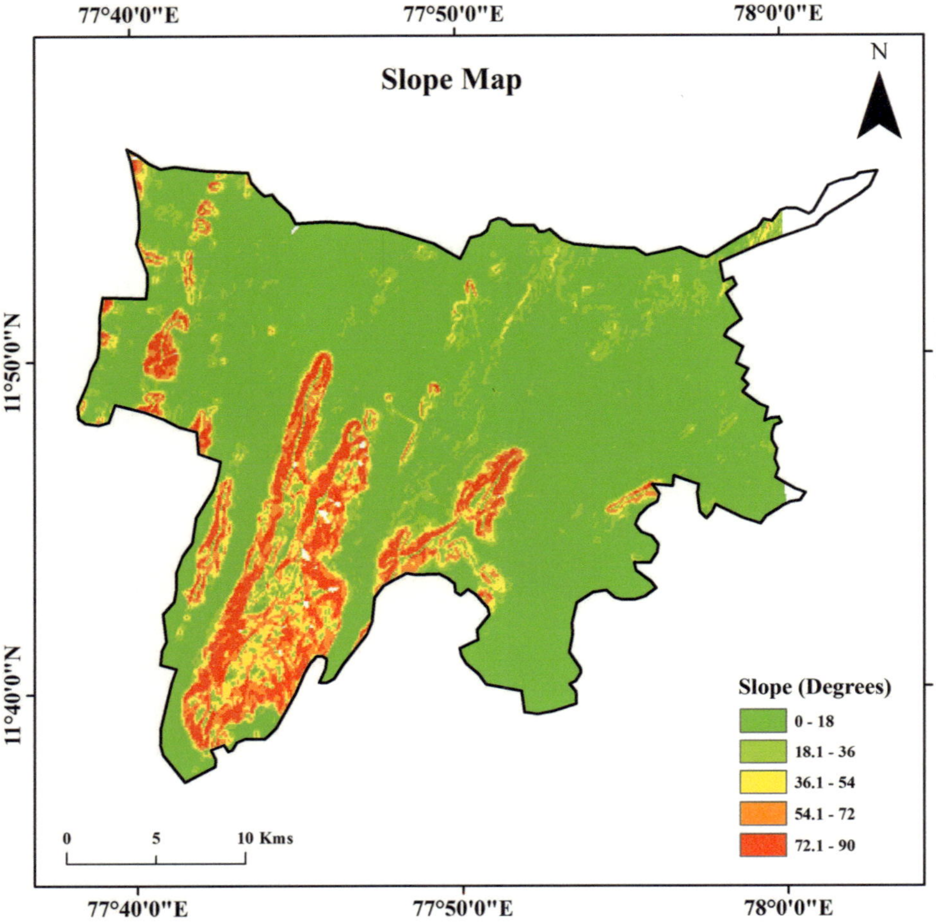

Figure 2.
Map showing slope in degrees of the study area.

3.2 Drainage

Drainage density is termed as the spacing distance of drainage (**Figure 3**). The ratio between the total length of the stream for each stream order. The drainage density (km/sq km) is categorized into five from 0 to 9100 sq/km. About 28% of the study area is having a drainage density of less than 840 km/sq km. The drainage density is inversely proportional to the porousness. The surface runoff will be more where the rainfall infiltration is less which happens in less porous soil. The weightage was decided accordingly with respect to the density/sq km ranging from 1 to 5.

3.3 Soil

The type of soil in the region reveals the groundwater recharge potential in the zone based on the soil properties such as permeability, conductivity etc., Based on these the weightage for the soil were given for sandy clay loam and loamy sand as 4, rock land as 1, sandy clay as 3, clay loam as 2 is shown in **Figure 4**. About 52% of the study area is of sandy loam followed by rocky land. The sandy loam soils have a high infiltration rate. More the infiltration rate, the more the groundwater potential.

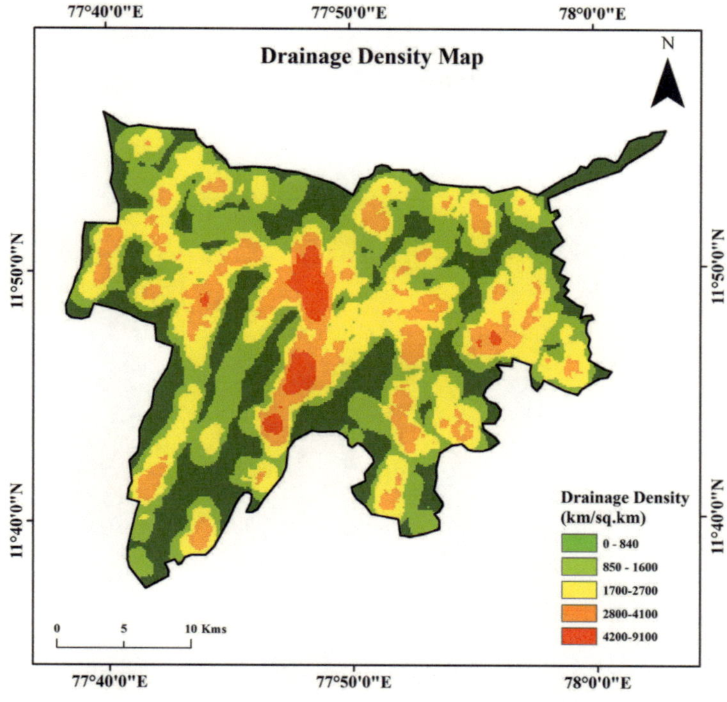

Figure 3.
Map showing drainage density in km/km² of the study area.

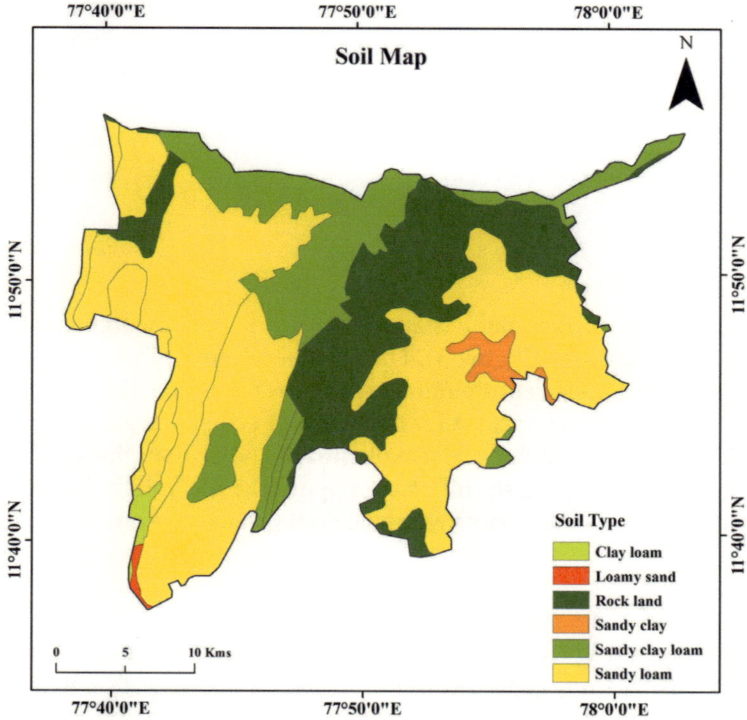

Figure 4.
Map showing soil type in the study area.

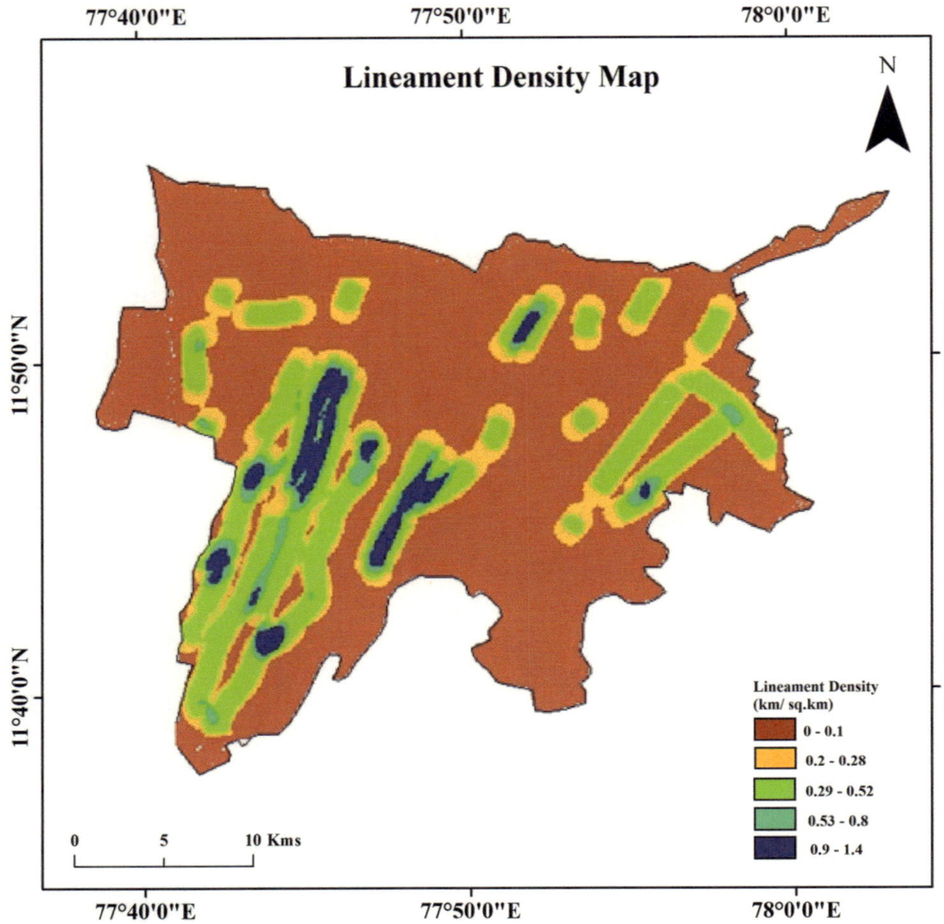

Figure 5.
Map showing lineament density in the study area.

3.4 Lineament

Lineaments are the surface manifestation of subsurface shortcoming or auxiliary relocation and deformations (**Figure 5**). The lineament density (km/sq km) varies from 0 to 1.4 km/sq km. It speaks to profound seated blames, breaks and joint sets, drainage lines and distinctive litho-contacts. It is a linear and curvilinear feature that is noteworthy for groundwater, mineral and metal explorations and exploitations. Lineaments were digitized after calculating the hill shade slope in the ASTERDEM data. Same as drainage density, lineament density's weightage was given based on their densities.

3.5 Geology

The geology type (**Figure 6**) in the study area has a more spread of crystalline rocks whose weightage was given as 1 based on its properties in conducting the groundwater recharge, followed by intrusive rocks having weightage same as prior. The semi consolidated sediments were given a weightage of 3 as it has moderate groundwater conducting properties.

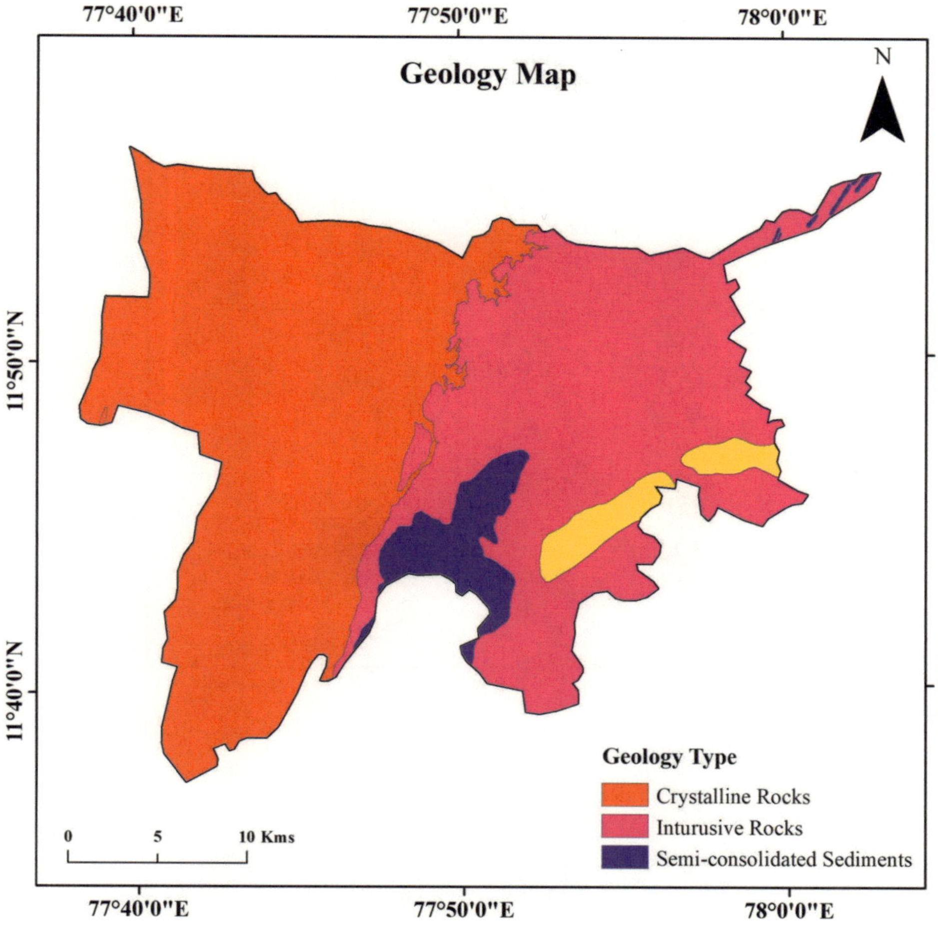

Figure 6.
Map showing geology type in the study area.

3.6 Geomorphology

The characteristics such as texture, tone, shape, color etc., are used for delineating the Geomorphologic units (**Figure 7**). Structural hills are observed in the study area, which mostly acts as runoff zones due to their sloping topography. The pediplain was given a weightage of 3, followed by, structured hills (2), denudational hills (2), flood plain (2), waterbody mask (5).

3.7 Landuse land cover

Baseline information about occurrences of surface and groundwater can be directly or indirectly obtained using land-use/ landcover information of that particular area. The LULC effect (**Figure 8**) can be calibrated by means of reducing runoff and facilitating, or by trapping water on their leaf. The study area is categorized into six land-use patterns as water bodies (5), dense forest (3), degraded forest (3), barren land (2), fallow land (4), agricultural land (4) from the Landsat 8 data and corresponding weightage were given during the analysis.

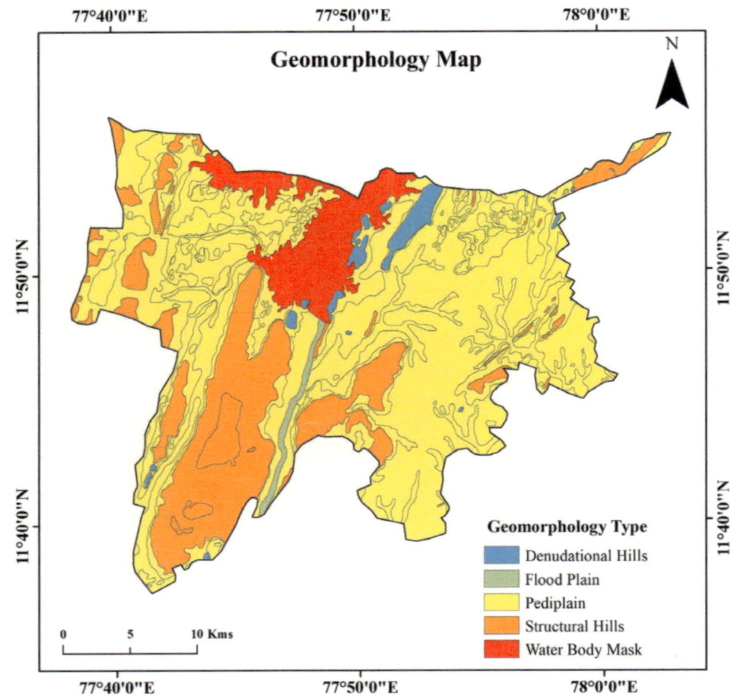

Figure 7.
Map showing geomorphology in the study area.

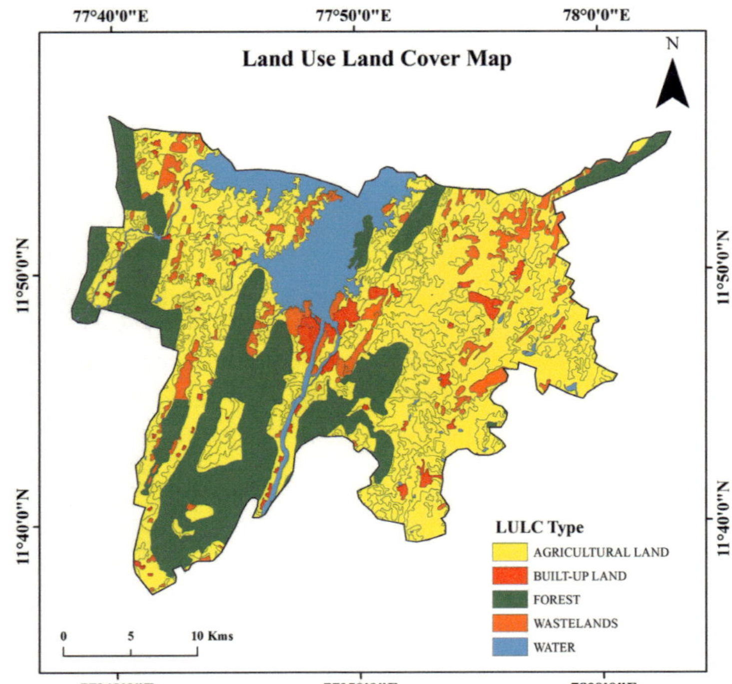

Figure 8.
Map showing LULC in the study area.

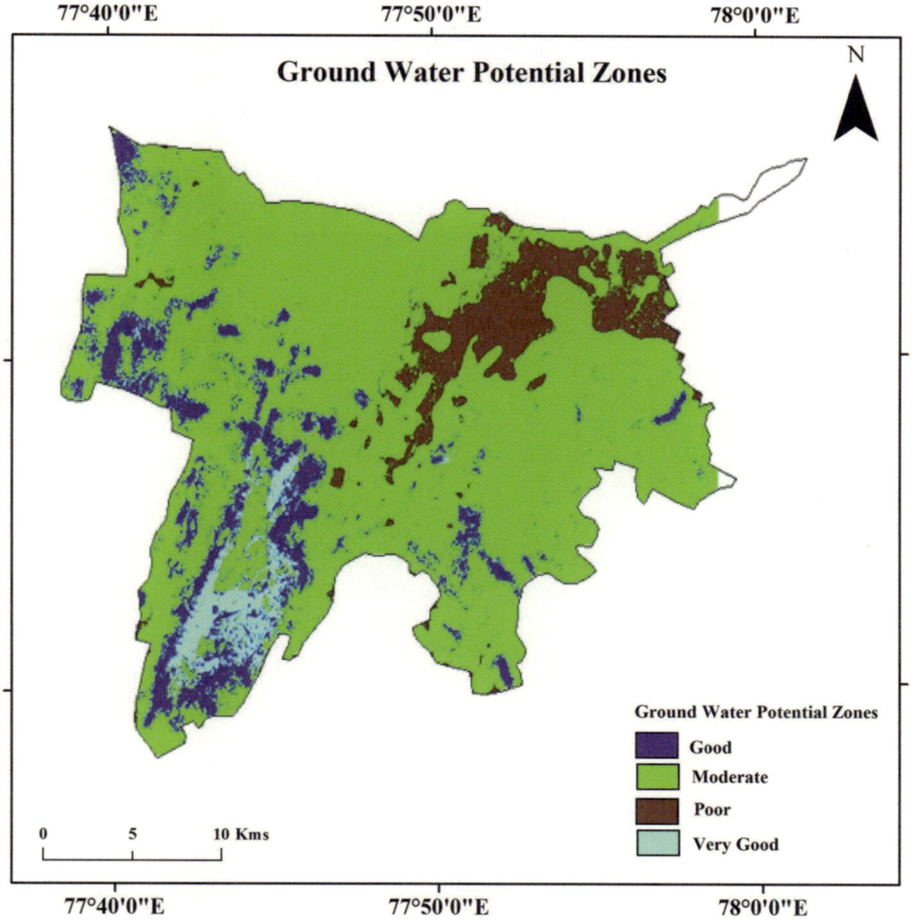

Figure 9.
Map showing groundwater potential zones in the study area.

3.8 Delineating the groundwater potential zone

The delineation of the groundwater potential zones has been created by means of integrating the thematic maps namely slope, drainage, soil, lithology, lineament, LULC, and rainfall using RS and GIS techniques. The interpreted layers were analyzed using weighted overlay method and resultant is the different potential zones (**Figure 9**). The south-west side of the study area is having a very good groundwater potential location compared to the north-east side. About 74.6% of the study area is having a moderate groundwater potential location, which can be utilized for artificial recharge of the wells and management plans.

4. Conclusion

The groundwater potential zone map has been generated from the various thematic layers based on their significance and corresponding classes. In the present research work six conditioning parameters namely (i) geology, (ii) slope,

(iii) geomorphology, (iv) land use/cover, (v) drainage and (vi) lineament were identified. These parameters were classified based on literature and expert opinion. The experts' knowledge was important to determine the rank of each conditioning factor. The parameters like lineament density, geology, elevation, slope etc. determine the groundwater potential significantly.

The groundwater potential zone of the study area varies from very good to poor. **Figure 8** shows the moderate level of potential zones is present in the study area due to various factors involved. About 74.6% of the total area falls under the.

"moderate" zone, 12.3% falls under "good" zone, 9.78% falls under "poor" zone, and 3.4% of the basin fall under "very good" zone. The results predict that if the groundwater potential zones were utilized properly for exploring the water resources, the drought can be coped up with potential agricultural practices.

Author details

C. Prakasam[1]* and R. Saravanan[2]

1 Department of Geography, School of Earth Sciences, Assam University (A Central University), Diphu, Karbi Anglong, Assam, India

2 Department of Civil Engineering, Chitkara University, Himachal Pradesh, India

*Address all correspondence to: cprakasam@gmail.com

References

[1] Magesh NS, Chandrasekar N, Soundranayagam JP. Delineation of groundwater potential zones in Theni district, Tamil Nadu, using remote sensing, GIS and MIF techniques. Geoscience Frontiers. 2012;**3**(2):189-196

[2] Waikar ML, Nilawar AP. Identification of groundwater potential zone using remote sensing and GIS technique. Int J Innov Res Sci Eng Technol. 2014;**3**(5):12163-12174

[3] Mwega BW, Mati B, Mulwa JK, Kituu GM. Identification of groundwater potential zones using remote sensing and GIS in Lake Chala watershed, Kenya. In: Mechanical Engineering Annual Conference on sustainable research and innovation. Thika, Kenya: Jomo Kenyatta University of Agriculture and Technology; 2013. pp. 42-46

[4] Ramu MB, Vinay M. Identification of ground water potential zones using GIS and remote sensing techniques: A case study of Mysore taluk, Karnataka. Int. J. Geomat. Geosci. 2014;**5**(3):393-403

[5] Rajendran S. Drought mitigation in Tamil Nadu. Economic and Political Weekly. 2014;**5**(3):393-403

[6] Owolabi ST, Madi K, Kalumba AM, Orimoloye IR. A groundwater potential zone mapping approach for semi-arid environments using remote sensing (RS), geographic information system (GIS), and analytical hierarchical process (AHP) techniques: A case study of Buffalo catchment, Eastern Cape, South Africa. Arabian Journal of Geosciences. 2020;**13**(22):1-17

[7] Doke AB, Zolekar RB, Patel H, Das S. Geospatial mapping of groundwater potential zones using multi-criteria decision-making AHP approach in a hardrock basaltic terrain in India. Ecological Indicators. 2021;**127**:107685

[8] Jhariya DC, Khan R, Mondal KC, Kumar T, Singh VK. Assessment of groundwater potential zone using GIS-based multi-influencing factor (MIF), multi-criteria decision analysis (MCDA) and electrical resistivity survey techniques in Raipur city, Chhattisgarh, India. AQUA—Water Infrastructure, Ecosystems and Society. 2021;**70**(3):375-400

[9] Melese T, Belay T. Groundwater potential zone mapping using analytical hierarchy process and gis in muga watershed. Abay Basin, Ethiopia: Global Challenges. 2022;**6**(1):2100068

[10] Saranya T, Saravanan S. Groundwater potential zone mapping using analytical hierarchy process (AHP) and GIS for Kancheepuram District, Tamilnadu, India. Modeling Earth Systems and Environment. 2020:1-18

Section 10

Traffic and Transport Management

Chapter 12

Optimization of the Road for Effective Management Traffic and Transport with GIS-GPS, Case Study: Pristina Capital

Arbnor Pajaziti and Orlat Tafilaj

Abstract

This chapter describes an integrated telegraphically GIS/GPS for an Urban Intelligent System and Traffic and Environmental Operations, where a telecommunications system enables the real-time reception of traffic data. We will focus on the improvement of bus services in Pristine capital including the implementation of more measures to improve travel time and reliability, providing new locations for "Park and Travel" and upgrade points were crossed from one form of transport to another, improving vehicle quality, providing better information and ticketing arrangement. Based on traffic surveys conducted of the Pristina Sustainable Urban Mobility Plan (PSUMP), new public transport bus routes have been designed to improve overall access/delivery services throughout the Pristine capital city. A new central infrastructure system traffic has also been proposed, aimed at improving the reliability and regularity of busses operating on these new lines. The new road configuration reflects current and future destinations, as well as passenger data obtained from public transport surveys, to help determine service frequencies and routes covered. The new bus network scheme is structured as the main and secondary/supplementary networks.

Keywords: GIS, GPS, integrated public transport system, mobility plan, bus routes, bus service line

1. Introduction

Geographic Information System is a data framework that has some expertise in the presentation, the board, investigation, and announcing of geographic data. GIS has changed and extended geology with your capacity to store a lot of information, break down them, and particularly present the consequences of customized map making. Among the wide scope of potential applications that GIS can utilize, transportation issues stand out enough to be noticed, **Figure 1**. One of the pieces of GIS that we are centered around the most is transport issues known as GIS-T.

The four fundamental parts of a GIS, coding, the board, investigation, and announcing, have explicit transport considerations [1].

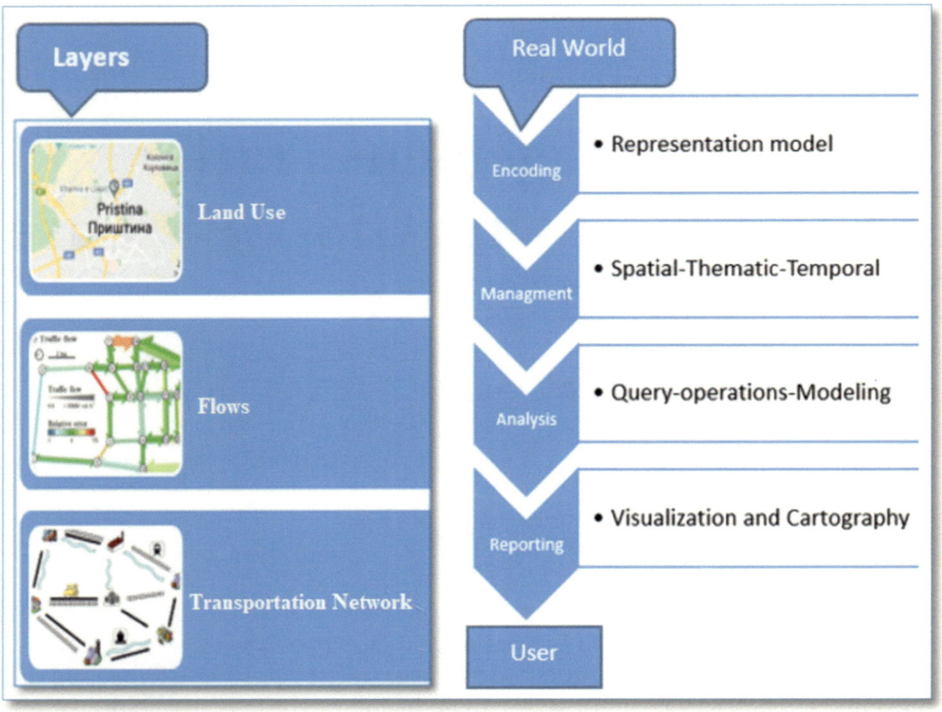

Figure 1.
Geographic information systems and transportation.

1.1 Coding

Coding means that a transport system to be used in GIS in the city of Prishtina must be coded. As an example of such coding in a road segment, we can take the number of lanes that segment has.

1.2 Management

Once the data is encrypted (e.g. the number of lanes in a road segment in the city of Prishtina) in this phase the storage of information in the database depends on the criteria of when it happened, where it happened, etc.

1.3 Analysis

Uses all available tools to analyze the situation in transport (e.g. the situation at peak hours on a road segment in Prishtina) and how much this situation will change in the case of new designs (e.g. the addition of another lane new, change of bus lines, etc.)

1.4 Reporting

GIS is completed with the ability to report the analyzed data. All information gathered from before stages are accounted for at this stage in visual structure (showed or printed) [2, 3].

Therefore, the data collected in this case by GIS-T will be understood much more simply in visual displayed forms (e.g. presentation of the largest number of accidents in an area of the city of Prishtina).

In the figure over, a vehicle framework is addressed as three layers connected with land use, releases (spatial cooperations), and organizations. Each has its attributes and related information It can be utilized freely or in the mix with different layers. Data System (GIS) is being utilized all the more generally in transport arranging offices, particularly among metropolitan vehicle associations. A lot more specialists are currently ready to involve GIS for thruways and transport the executives, because of falling expenses and expanded utilization of GIS. GIS gives transport a mechanism for putting away and breaking down information on land use, and travel conduct. The main targets for GIS utilization are planning/show and information integration.

Organizations need to distinguish potential issues that can be tended to through a GIS application all the more proficiently and actually, and more monetarily than with winning strategies [4, 5].

2. Optimization of the road for effective traffic management and transportation by GPS-GIS in the municipality of Pristina

Today we are generally seeing tremendous improvement in course and transport, both actually and mechanically. The entire advancement of these circles of public activity surprisingly is progressively coordinated towards the productive and viable utilization of restricted assets, whether material or monetary. To make the most of the transport branches so that they are efficient and effective, it is necessary to optimize the transport routes. To dig into the transport branches so they are proficient and viable, it is important to enhance the transport courses.

We will focus on improving bus services in Pristina including the implementation of more measures to improve travel time and reliability, the provision of new "Park and Travel" places (see the section on parking), and the improvement of points was switching from one form of transport to another, improving the quality of vehicles, providing better information and arranging to the ticket.

It is suggested that the degree of the purpose of public transport can be expanded through the new "main" organization of transports that will give quicker and more regular help. Then again, this will attract a larger number of users, coupled with a new design approach to the main locations where modes of transport can be changed ("Transfer points" - "Hubs") to create additional capacity and ease of movement for passengers, with minimal delay between services. The so-called "secondary" services are recommended, which will operate at a lower frequency than the main services but will provide important connections to residential areas, suburbs, and villages [6].

2.1 Summary of problems in Pristina

2.1.1 Fragmented bus service network

In the city, many bus services operate on the same roads as long as there are areas where the services do not work. Currently, local bus services are not coordinated in terms of fares or timetables, and passengers who need to travel to their destination pay more, spend more time traveling, and/or have to walk part of the way. A more cohesive system would bring benefits to both operators if implemented systematically.

2.1.2 Reliability of bus services

The increase in congestion levels in traffic, especially during peak periods of morning and evening travel, creates problems for bus services in terms of the ability to travel on time according to the planned schedule. This uncertainty makes it difficult for travelers to plan their trips to school, work, or shopping, and sometimes people waiting at bus stops are unsure of when the next bus will arrive.

2.1.3 The need for better information on bus services

For many locals, employees, and visitors to Pristina, it is difficult to understand the current network of busses and timetables, with different services and different destinations.

Similarly, people waiting in place for bus stops are unsure when the busses will arrive. Improving information will increase people's confidence in using the services.

2.1.4 Request for regulation of taxi services

Although several good quality taxi operators are operating in Pristina, there are also a large number of illegal taxis, which cause problems in terms of competition with other local operators as well as bus operators. It is important to regulate, monitor, and control taxi services effectively to ensure high quality and safe passenger service.

Based on the research undertaken to develop the SMUPP (Sustainable Mobility Urban Plan of Pristina), several important issues and problems related to public transport in the city have been identified, including those listed below:

- Many busses are old and do not meet the specific essentials and guidelines expected for present-day traveler transport as far as access and outflow levels (motor principles), and standards required for modern passenger transport in terms of access and emission levels (engine standards);

- Some areas do not have transport lines (or have a limited level of service), with some roads too narrow to accommodate standard busses;

- Uncoordinated timetables and non-integration of bus services and lines throughout the city, reduce opportunities for connecting services for passengers;

- The ticketing system is unattractive and not well integrated, where for each route/service a separate ticket is required;

- Missing data and information regarding bus service/stopping details including name, signs, and information (mainly timetables) at bus stops;

- Many public transport documents and data provide different and contradictory information;

- There is no complete overview of routes and bus operators, number of bus lines, bus route itineraries, with incomplete public transport data available;
- There is also a lack of data on public transport operators regarding the use of busses/services, as well as equipment at the stop, etc.; and,
- Many services operate in similar ways and there is evidence of service duplication that requires special attention in the context of the overall review and restructuring of the city network.

2.2 Proposed strategy: new bus network throughout the city

Based on traffic surveys conducted by the SMUPP, new public transport bus routes have been designed to improve access/overall service delivery throughout the city. A new central traffic infrastructure system has also been proposed that aims improve the reliability and regularity of busses operating on these new routes. The new network approach is designed to respond to changing demands for public transport in the city, ensuring good coverage, and reducing overlap and duplication of some current lines. The new road configuration reflects current and future surface destinations, as well as passenger data taken from public transport surveys, to help determine service frequencies and roads covered. The new transport network plot is organized as follows:

Main network: based on the desire to increase the coverage and frequency of bus services as an attractive alternative to the use of vehicles for downtown travel. One of the principal goals of these administrations is to provide the most reliable speeds and operating schedules to assist in attracting new passengers. The main network includes the current bus service lines 1–7, **Figures 2–6**.

Secondary/supplementary network: In support of the main network, several secondary or supplementary roads have been designed that would connect the main urban center with the peripheral (residential) areas, as well as beyond. These secondary services would also connect more rural and suburban areas to the main urban center. The supplementary network includes bus service lines number 10–21 and 2A, 2B, and 2C.

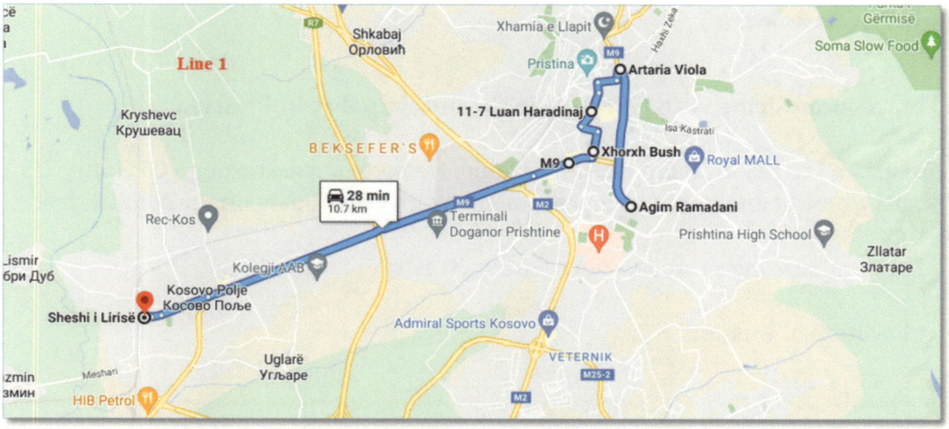

Figure 2.
Line 1 (Technical Faculty—Fushë Kosovë) [6].

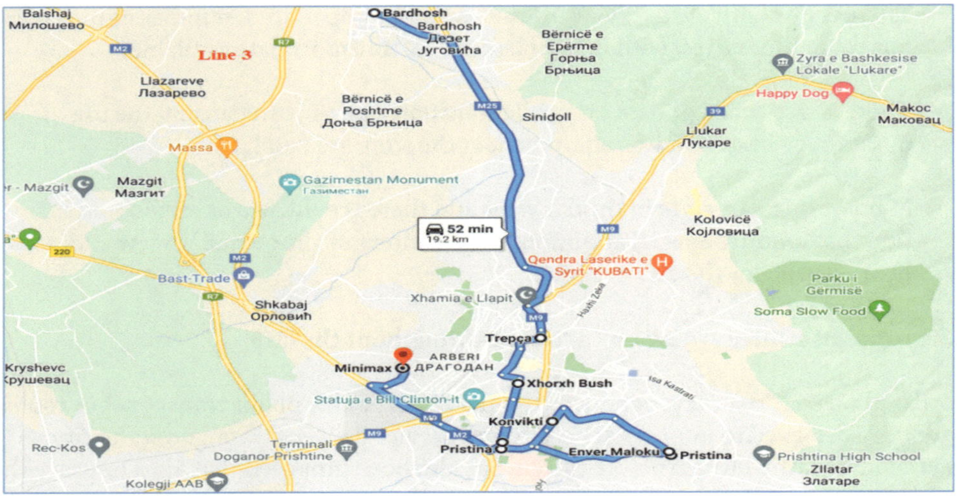

Figure 3.
Line 3 (Mati District—Bardhosh Village).

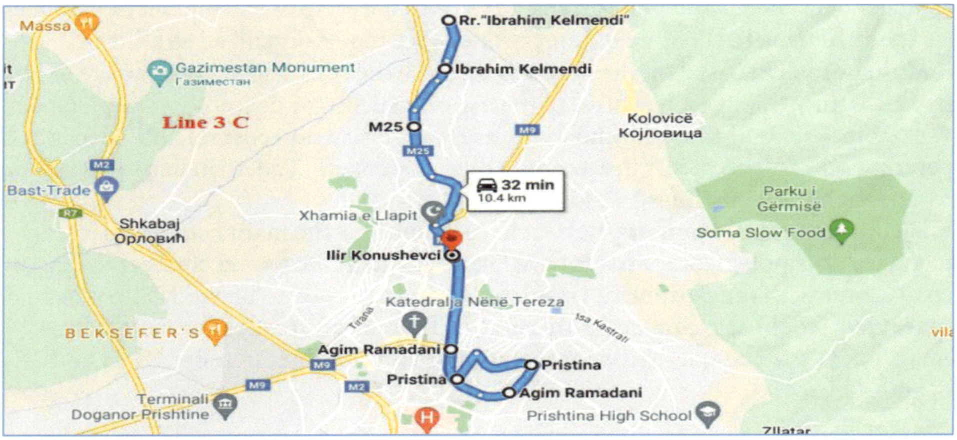

Figure 4.
Line 3C (Sinidoll—Street "B").

2.3 Local monitoring with GPS-GIS in the municipality of Pristina

In urban traffic, local monitoring is done, tracking busses through GPS and GIS systems where at any time and any moment are given real data on the condition of busses.

These centers monitored various data such as:

- Location of busses;
- Speed;
- Fuel level;

Figure 5.
Line 15 (Keqekollë Village).

Figure 6.
Line 7C (Kalabri - economic Center - Bus Station - Kalabri).

- Condition of busses (stopped - on);
- The trajectory of the last road;
- History of busses (data stored for a longer period, if necessary, date search), etc.

The monitoring center presents the current condition of all busses as shown in **Figures 7** and **8**.

The program also realizes the demand for our needs depending on what and when we want to monitor. We can get different settings such as:

- Driver details;

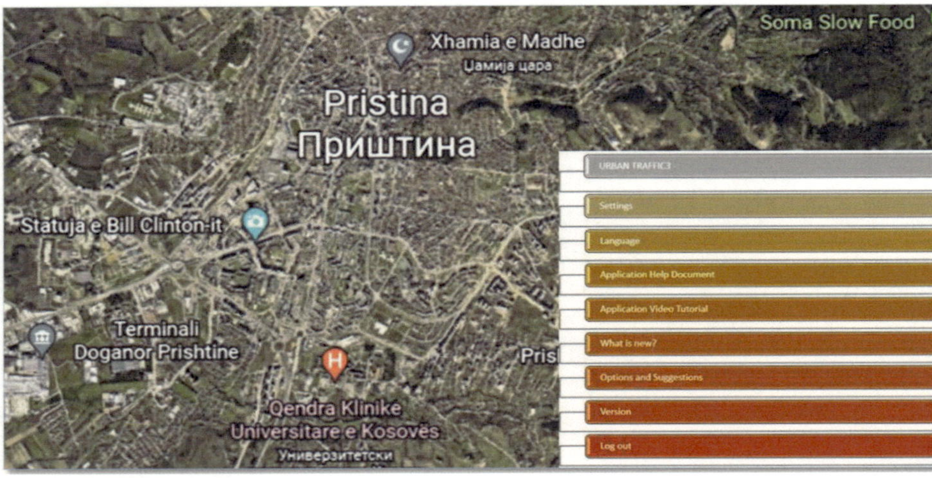

Figure 7.
Management by urban traffic software.

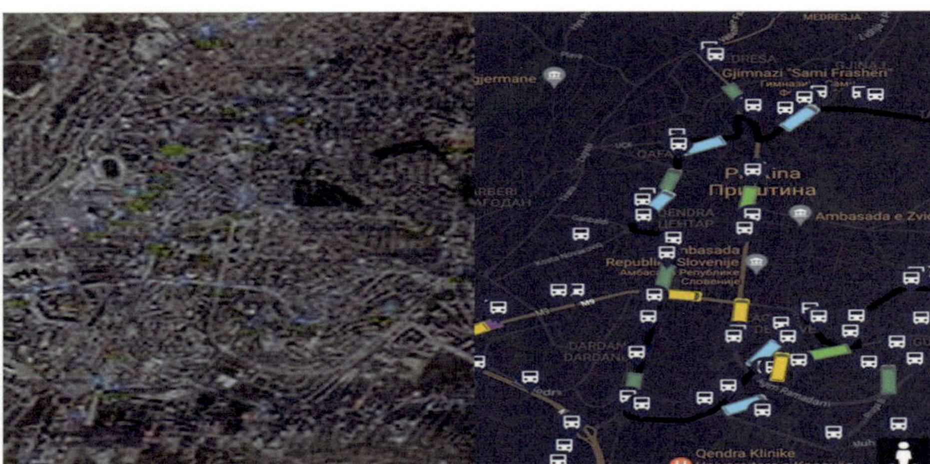

Figure 8.
Cartographic representation of bus location.

- Data for the bus (make, model, daily distance, speed, GPS number of the GPS device; license plates, the fuel level in liters and in percentage, the period when the vehicle was turned on or stopped), etc.

The introduction of information should be possible through the different diagrams given by the product as displayed in **Figure 9**.

The following case shows the monitoring of fuel condition and speed where the presentation can be made for one bus or all available busses, **Figure 10**.

In many cases, the company needs to search the history of busses where this data is stored with great accuracy. In the following case, the search for the history of a bus is explained, as well as the presentation of these data, as well as the possibility of viewing the route, traveled for the mentioned date through animation.

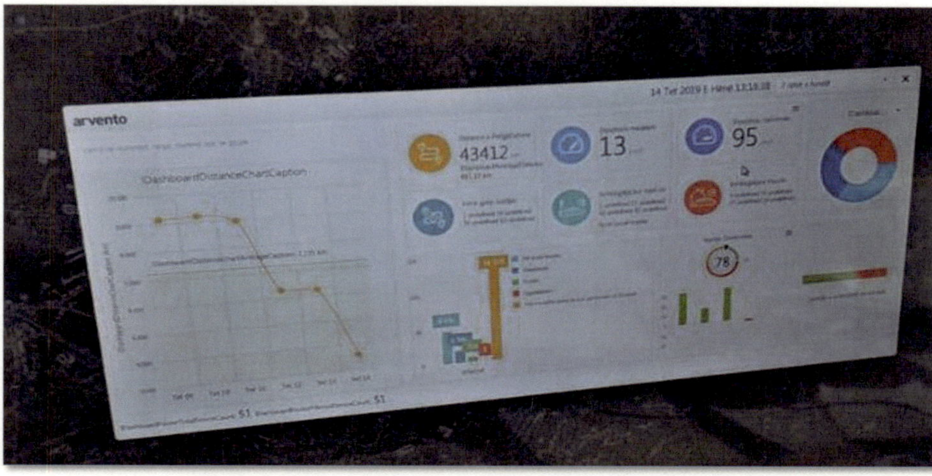

Figure 9.
Show information through charts.

Figure 10.
Monitoring the fuel condition of a bus.

3. ARCGIS support package and platform

ArcGIS is a framework that empowers the formation of information, guides, and models to be utilized by clients in a work area, program, or through gadgets like cell phones, GPS, and so forth, contingent upon the requirements of the association.

People use ArcGIS in all types of organizations, to improve their workflow and resolve critical situations that arise. ArcGIS helps to:

- Data and value management (including management of the territory, services, customers, etc.).

- Planning and analysis (includes forecasts and risk analysis in the organization). Business operations (distribution, monitoring, tracking, inspection, maintenance, etc.).

3.1 Decision-making

Below (**Figure 11**) is the complete scheme of an ArcGIS platform. The presentation is made according to two levels, the functional one and the application one. ArcGIS components and their component applications are presented.

ArcGIS is a set of software products integrated that serve to set up a complete ArcGIS system. It consists of 4 basic structures [7]:

ArcGIS Desktop—It is a gathering of GIS applications comprising 3 primary items ArcView, ArcEditor, and ArcInfo.

GIS Server—ArcIMS, ArcGIS Server, and ArcGIS Image Server. They are all products that support their online functionalities. Used in the distribution of dynamic maps and various GIS services through the Internet.

Mobile GIS—Mobile GIS devices, such as GPS, assist staff members who do not have extensive experience in using GIS applications.

ESRI Developer(EDN)—This contains integrated software components for GIS developers (programmers) that help build GIS applications and provide optimal solutions to various problems. Provides online code resources, reference libraries, object diagram models, and technical support.

3.2 GIS web platform logic

To better understand what the platform is based on, we need to follow step by step the logic on which it is built.

In the beginning, we will present concisely the general scheme in which the data circulate as well as the environment in which it works, **Figure 12**. As should be visible from the figure, the framework executives supply the application interface in two equal structures to one another. On the one hand, we have geographical data which directly affects the appearance on the map, while on the other hand, we have maps dynamics generated through utility packages such as ArcMap, ArcCatalog, etc. which generate maps and statistics which are structured according to the respective categories and are required based on filtering [7, 8].

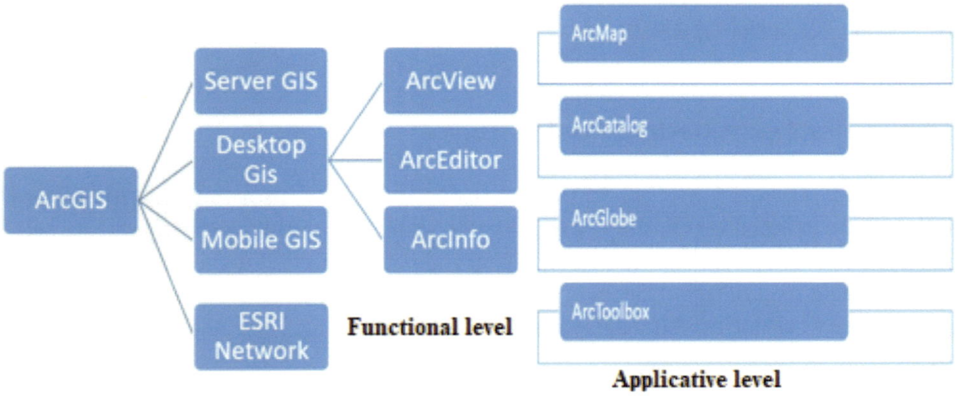

Figure 11.
ArcGIS helper platform.

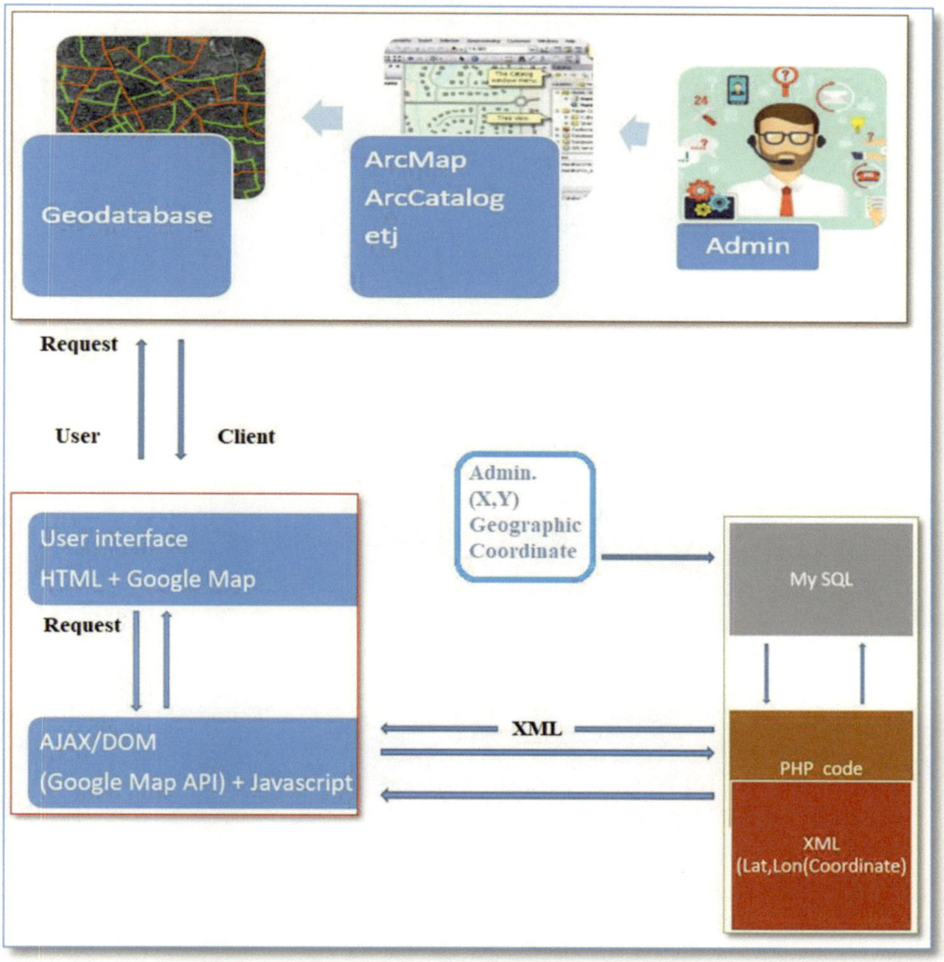

Figure 12.
The logic of the GIS web platform.

3.3 Practical case - data processing from GPS device and arc map software

GPS device enables us to obtain data respectively locations, one of them is also receiving in Excel and Gpx format. In the following case, we will present the processing of GPS data from gpx format via Arc Map. Data storage from GPS equipment is done directly in Gpx format where data insertion and processing are done automatically by Arc Map software. In case of data entry (location points the past route of the vehicle will be displayed). **Figures 13–15** show the case of gpx data insertion.

4. Conclusion

In this chapter, we described the optimization of transport routes through GPS / GIS technology. Today we are increasingly living in a society where most decision-making is based on geographic information. Maps are turning into a defining element

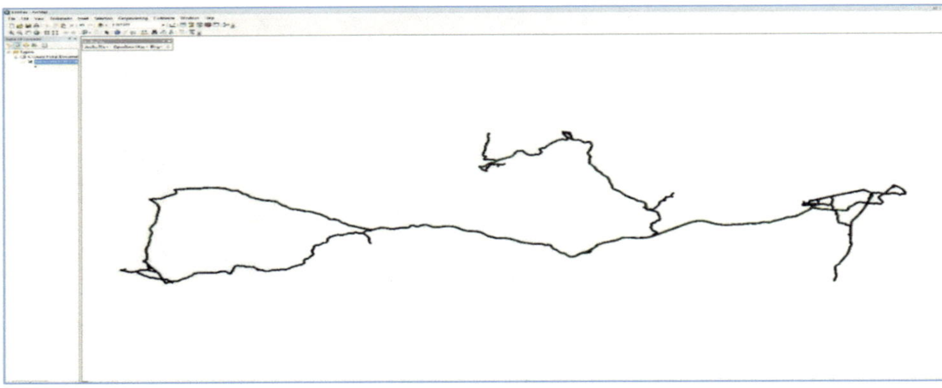

Figure 13.
Creating the route from GPS data points in arc map [8].

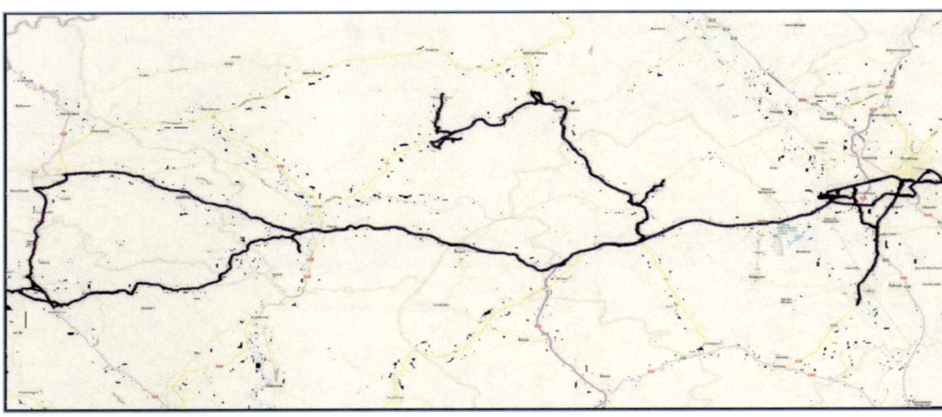

Figure 14.
Inserting the map [8].

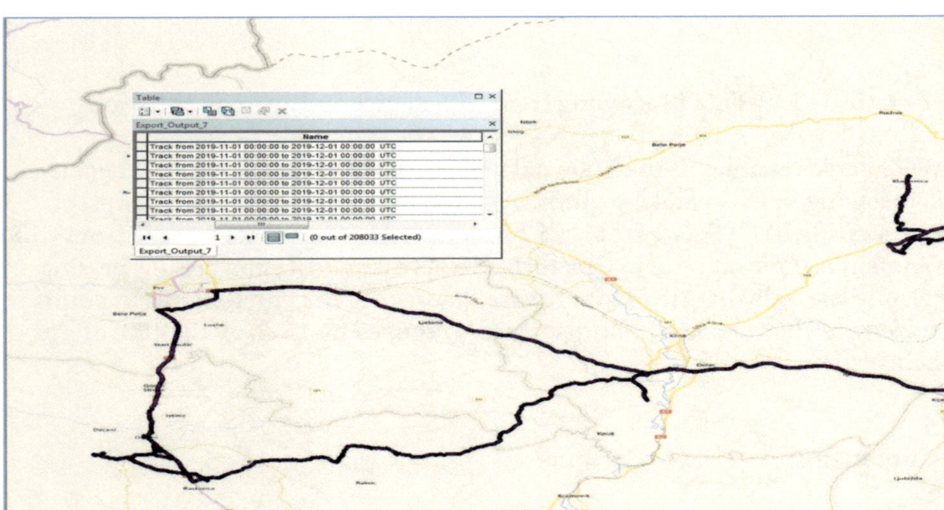

Figure 15.
GPS data display, attribute table [8].

by translating numbers and figures into visuals many times easier to interpret and analyze. As in many other countries of the world, in Kosovo, web GIS applications are slowly finding their space in competitive, educational, informational, environments, etc. However, this process requires greater support from state and private entities through funding and incentives for economic and formative aspects. Experience shows that many countries already have integral and inseparable parts of GIS systems in almost all information and communication sectors. Based on the events of recent years that have occurred in Kosovo, such as tragic traffic accidents, major traffic jams, illegal construction, educational elements and science, poorly managed projects and platforms, it is intended to study the possibility and potential of integrating this technology as the optimal solution for addressing these issues so delicate to the environment.

As a conclusion of the above quotations, it is intended to build a multifunctional platform with open access for all people (open-access). In this platform, there will be several state and independent institutions and organizations, but which will be open to any entity that provides information is checked and verified for their consistency and accuracy, allowed to be integrated as part (product) of questions for the GIS platform.

The proposed system offers opportunities for cost-effective expansion of traffic and environmental data coverage, being a promising tool for the municipality, usually characterized by large budget constraints. To improve system analysis skills and provide more relevant information to citizens, traffic simulations and environmental impact should be included. GPS/GIS combination is perhaps the most conspicuous area of examination in its field.

A GIS-based application was selected as the best alternative to improve accuracy and timeliness in the priorities of traffic delays, faster roads, accident location, etc. In terms of future work, the system can be upgraded to provide more features by following these ways:

- Add multimedia capabilities such as the ability to store digital photos.
- Addition of printing function that can print accident information.
- Addition of a function that can capture the accident scheme.
- Changing the database from text format to tabular format.
- Collaboration by the person who is directly or indirectly involved.
- Cooperation from Urban Traffic and Accident Registration Center (Kosovo Police).

Author details

Arbnor Pajaziti and Orlat Tafilaj*
Faculty of Mechanical Engineering, University of Prishtina, Kosovo

*Address all correspondence to: orlattafilaj@gmail.com

References

[1] https://transportgeography.org/contents/methods/methods-transport-geography/transportation-gis [Last update 7 March 2022]

[2] Longley P. Transportation GIS: GIS-T. In: Longley P, Goodchild M, Maguire D, Rhind D, editors. Geographical Information Systems: Principles, Techniques, Applications, and Management. New York: Wiley; 1999. pp. 827-844

[3] https://transportgeography.org/contents/methods/geographic-information-systems-transportation/

[4] Rodrigue J-P. The Geography of Transport Systems. 5th ed. New York: Routledge. p. 456

[5] Pajaziti A. Intelligent Transportation Systems. Prishtina: University of Prishtina; 2020

[6] https://www.studymode.com/essays/Supply-Chain-Management-887219.html [Last update 9 March 2022]

[7] Hysenaj M. Geographical Information Systems. Shkodër, Albania; 2011

[8] Tafilaj O. Advanced Vehicle Driver Support Systems. Prishtina: University of Prishtina; 2013